ISBN 978-1-333-23982-4
PIBN 10477915

1 MONTH OF
FREE
READING

at

www.ForgottenBooks.com

By purchasing this book you are eligible for one month membership to ForgottenBooks.com, giving you unlimited access to our entire collection of over 700,000 titles via our web site and mobile apps.

To claim your free month visit:

www.forgottenbooks.com/free477915

English
Français
Deutsche
Italiano
Español
Português

www.forgottenbooks.com

Mythology Photography **Fiction**
Fishing Christianity **Art** Cooking
Essays Buddhism Freemasonry
Medicine **Biology** Music **Ancient
Egypt** Evolution Carpentry Physics
Dance Geology **Mathematics** Fitness
Shakespeare **Folklore** Yoga Marketing
Confidence Immortality Biographies
Poetry **Psychology** Witchcraft
Electronics Chemistry History **Law**
Accounting **Philosophy** Anthropology
Alchemy Drama Quantum Mechanics
Atheism Sexual Health **Ancient History**
Entrepreneurship Languages Sport
Paleontology Needlework Islam
Metaphysics Investment Archaeology
Parenting Statistics Criminology
Motivational

THE

JOURNAL

OF THE

BOMBAY NATURAL HISTORY SOCIETY.

EDITED BY

R. A. STERNDALE

and

E. H. AITKEN.

VOLUME I.

1886.

Consisting of Four Numbers and containing Twelve Illustrations.

𝕭𝖔𝖒𝖇𝖆𝖞:

PRINTED AT THE

"CAXTON STEAM PRINTING WORKS," BOMBAY.

1886.

CONTENTS.

PAGE.

JOURNAL

OF THE

BOMBAY

Natural History Society.

No. 1.]   BOMBAY, JANUARY 1886.   [Vol. I.

INTRODUCTION.

In issuing the first number of the Journal of the Bombay Natural History Society, it seems appropriate to give some account of the origin of the Society and of its position at the present time. It was founded on the 15th of September 1883, by seven gentlemen interested in natural history, who proposed to meet monthly and exchange notes, exhibit interesting specimens, and otherwise encourage one another. The subscription was purposely made little more than nominal, and the possibility of forming or maintaining a museum was scarcely contemplated at that time. For several months meetings were held in the Victoria and Albert Museum; but in January 1884, Mr. H. M. Phipson kindly offered the use of a room in his office in the Fort. This removal to a central situation gave an astonishing impulse to the Society. The meetings were better attended, the membership increased, and collections began to be made, so that in a very short time the necessity for more ample accommodation was pressingly felt. A committee appointed to seek for suitable rooms, having failed elsewhere, recommended the Society to ask Mr. Phipson to let one half of his office premises, including the room, of which they had up to this time had the gratuitous use. He consented to this, and so the Society continued to hold its meetings and keep its collections at 18, Forbes Street. Its progress was so rapid, however, that these premises were soon felt to be too small, and last month the collections were removed to larger and in every way more suitable rooms at 6, Apollo Street.

1

In the month of May last, a very important change was made in the constitution of the Society. The monthly meetings, being largely attended by members who took only a general interest in natural history, had naturally acquired a very popular character, and it was found impossible to introduce much strictly scientific matter on these occasions. It was decided, therefore, while continuing the good work already done in popularising the subject of natural history, to make better provision for the study and advancement of the science by the formation of sections as follows :—

1. Mammals and Birds.
2. Reptiles and Fishes.
3. Insects.
4. Other Invertebrata.
5. Botany.

Those members specially interested in any branch of natural history were invited to join the corresponding section, elect a President and Secretary, take charge of that portion of the collections which appertained to their division, and hold their own meetings, thus forming practically a separate Society affiliated to the general Society. All the sections have now organised themselves and made their own arrangements for carrying on the work of their respective branches.

This has prepared the way for another important and necessary step—the publication of a journal in which whatever of value or interest is transacted at the sectional meetings may be permanently put on record. Till now there has been no publication in the Bombay Presidency devoted to natural history or any of its branches, and, perhaps as a consequence of this, there is scarcely any Presidency or Province the fauna of which has received so little attention. It is hoped that the introduction of this journal will stimulate lovers of Nature, especially in all parts of this Presidency, to record and communicate their observations. In accordance with the character which this Society has assumed from the beginning, the aim of its journal will be, as far as possible, to interest all students of Nature, ever remembering that there are many Naturalists, in the highest sense of the term, who have not such a technical knowledge of any particular branch of the science as to be able to enter with interest into questions of nomenclature and the discrimination of closely allied species. The Secretaries of the Sections would therefore invite sportsmen and others to communicate anything interesting or worthy of note, which comes under their observation, bearing on the nature and habits of animals or plants.

One other matter remains to be noticed. No public library in Bombay affords much assistance to the naturalist, and the absolute necessity of having a good library of their own early forced itself on the attention of the members of this Society. Unfortunately it is impossible to set aside any adequate sum out of the ordinary income of the Society for the purpose; but on two occasions special subscriptions have been collected and a number of valuable and necessary works secured, while several members have at different times contributed useful books. The Bombay Government has also been so good as to present the Society with all the parts of Sir J. Hooker's Indian Flora already issued and to promise the rest. With all this, however, it is impossible to say more than that a beginning has been made, but as the usefulness of the Society continues to be day by day more widely appreciated, the committee look confidently forward to the time when an adequate Library Fund will be formed and maintained by spontaneous donations.

LIST OF MEMBERS.

President.

H. E. the Right Honorable Lord Reay, C.I.E., LL.D., F.R.G.S.

Vice-Presidents.

Dr. G. A. Maconachie, M.D., C.M.
Dr. D. MacDonald, M.D., B.S.C., C.M.
The Hon'ble Justice Birdwood, M.A., LL.M. (Cantab.)

Treasurer and Secretary.

Mr. E. H. Aitken.

1st Section.—(*Mammals and Birds.*)

President—Mr. R. A. Sterndale, F.R. G.S., F.Z.S.
Secretary—Mr. E. H. Aitken.

2nd Section.—(*Reptiles and Fishes.*)

President—Mr. G. W. Vidal, C.S.
Secretary—Mr. H. M. Phipson.

3rd Section.—(*Insects.*)

President—*Vacant.*
Secretary—Mr. E. H. Aitken.

4th Section.—(*Other Invertebrata.*)

President—Dr. G. A. Maconachie, M.D., C.M.
Secretary—Mr. J. C. Anderson.

5th Section.—(*Botany.*)

President—The Hon'ble Justice Birdwood, M.A., LL.M. (Cantab.)
Secretary—Surgeon K. R. Kirtikar, F.S.M. (France), F.R.C.S.

NAME.	RESIDENCE.
Adams, J. B. D.	...Tanna
Aitken, D. E.	...Kurrachee
Aitken, E. H.	...Bombay
Almon, W.	...Bombay
Anderson, J. C.	...Bombay
Anderson, W. C.	...Bombay
Anderson, Capt. W. R.	...Simla
Arnott, Surgeon-Major J.	...Bombay
Arthur, A.	...Bombay
Baddeley, Lieut.-Colonel	...Cawnpore
Bainbridge, Surgeon-Major G.	...Bombay
Baines, J. A. (C.S.)	...Dharwar
Baker, W. A.	.. Bombay
Bankier, W. A.	...Bombay
Barnes, D. E. H.	...Saugor
Barton, E. L.	...Bombay
Bashford, T.	...Bombay
Baumbach, R.	...Bombay
Beardmore, E. B.	...Badnera
Becher, Captain (F.Z.A.)	...Kirkee
Begbie, J. A.	...Bombay
Bell, T. R. D.	...Canara
Best, W. J.	...Bombay
Betham, W. G.	...Nasik
Beyts, C. A.	...Bombay
Bicknell, H.	...Bombay
Bingham, Captain C. H.	...British Burma
Birdwood, the Hon'ble Justice	...Bombay
Blackwell, G. F.	...Tanna
Blackwell, H. F.	...Bombay
Blum, M.	...Bombay
Bousted, Surgeon-Major	...*Europe*
Branson, R. M.	...Bombay
Brendon, S.	...Bombay
Brown, J. W.	...Bombay
Brunton, R. P.	...Bombay
Buckland, H. W.	...Bombay
Bullock, W.	...Bandora
Butcher, L. H.	———
Byrne, C. H.	...Bombay

Name.	Residence.
Calthrop, E.	...Bombay
Campbell, E. W.	...Bombay
Campbell, John	...Bombay
Cane, Rev. A. G.	...Poona
Cassels, J. A.	...Bandora
Carroll, E. B.	...Bombay
Chatfield, K. M.	...Poona
Christian, A. G.	...Bombay
Clark, Captain A. E.	...Bombay
Close, E. P.	...Bombay
Coles, Colonel	...Bombay
Collie, Surgeon R.	...Bombay
Collister, J. G. H.	...Bombay
Conder, H.	...Bombay
Coussmaker, Major M. F.	*...Europe*
Cooke, Dr. T.	...Poona
Cooper, C. P.	...Bombay
Cornforth, J. P.	...Bombay
Cox, A. F.	...Bombay
Crawford, L.	...Bombay
Crawley-Boevey, A. W. (c.s.)	...Ahmedabad
Creagh, Captain R. P.	...Bombay
Crockett, Captain	...Bombay
Cuffe, T. W.	...Bombay
Curjel, H.	...Bombay
Cursetjee, Jehangir Maneckjee	...Ratnagiri
Curwen, H.	...Bombay
Daver, Framjee N.	...Bombay
Davidson, J. (c.s.)	...Mallegaum
Dewar, Miss	...Bombay
Ditmas, A. R.	...Vellore
Doig, S.	...Ahmedabad
Dreckman, Rev. F.	...Bombay
Duigan, S. A.	...Ajmere
Dubash, Sorabjee D.	...Bombay
Dumayne, F. G.	...Bombay
Dumbell, H. C.	...Bombay
Dymock, Dr.	...Bombay
Edgelow, F.	...Bombay
Ellaby, Miss C. (m.d.)	...Bombay
Fairbank, Rev. S.	...Ahmednugger
Ferguson, Dr. A. F.	...Bombay
Ffennell, Rev. W. J.	...Bombay
Forbes, Mrs. C. H. B.	...Bombay
Forrest, G. W.	...Poona
Forrest, L. R. W.	...Bombay
Francke, A.	...Bombay

NAME.	RESIDENCE.
Gaddum, F. ...	...Bombay
Gay, E.	...Calcutta
Gibson, G. ...	...Khandesh
Gilbert, R. ...	...Bombay
Gleadow, W.	...Sukkur
Goldsmid, F.	...Belgaum
Gopal, Babajee	...Bombay
Gordon, H. K.	...Bombay
Graham, W. D.	...Bombay
Grant, G. F. M.	...Tanna
Gray, A.	...Bombay
Gray, C.	...Bombay
Gray, Dr. Temperley	...Bombay
Gray, Dr. Wellington	...*Europe*
Greaves, W.	...Bombay
Grieve, Rev. A. C. ...	...Bombay
Griffiths, J. ...	...Bombay
Gunthorpe, Lieut.-Colonel	...Berar
Gwyn, Captain A....	...Bombay
Hatch, H. F.	...Ratnagiri
Hatch, Dr. ...	...Bombay
Hart, Mrs. W. E. ...	...Bombay
Hill, F. J. A.	...England
Hore, Fraser S.	...Bombay
Inverarity, J. D.	...Bombay
Jackson, W. L. C. ...	...Bombay
Jacob, H. P.	...*Europe*
Jefferson, F. G.	...Bombay
Jefferson, J....	...Bombay
Jervis, W. H.	...Bombay
Johnson, J. R. Kirby	...Bombay
Johnstone, Miss	...Bombay
Jordan, E. V.	...Bombay
Kama, K. R.	...Bombay
Kama, Rustom K. R.	...Bombay
Kane, C. E....	...Bombay
Kanga, Dinsha P.	...Bombay
Ker, L. B. ...	...Bombay
Kiddle, S....	...Bombay
King, Alfred	...Bombay
Kingsley, F. G.	...Bombay
Kirtikar, Surgeon K. R. ...	...Tanna
Lang, T.	...Bombay
Langley, Surgeon-Major	...Bombay
Latham, the Hon'ble Mr. F. L.	...Bombay
Lester, C. F. G.	...Kathiawar

NAME.	RESIDENCE.
Lidbetter, T	...Bombay
Little, F. A.	...Bombay
Littledale, H.	...Baroda
Lynch, C. P.	...Bombay
Lynch, M. P.	...Bombay
MacCartie, Dr.	...Bombay
Macdonald, Dr. D. ...	...Bombay
Macdonald, J.	...Bombay
Macdonald, W. M. ...	...Bombay
Maconachie, Dr. G. A.	...Bombay
Mactaggart, H. M. ...	...Bombay
Marriott, F. W.	...———
McCann, W. H.	...Bombay
McEwen, R. ...	...Bombay
Mercer, F. ...	...Bombay
Messent, P. G.	...Bombay
Miller, E. ...	...Bombay
Minter, Capt. J. S.	...Bombay
Montgomery, T. R. A. G. ...	...Bombay
Morris, J. ...	...Bombay
Moscardi, E. S. (c s.)	...*Europe*
Mukund Ragoonath	...Bombay
Murphy, Dr. ...	...Bombay
Murray, G. S.	...Bombay
Murray, W. ...	...Bombay
Murray, J. A.	...Kurrachee
Naigaumwalla, K. D.	...Bombay
Naoroji, Rev. Dhanjibhai ...	...Bombay
Newborn, C.	...Bombay
Newnham, A.	...Bhuj
Nicholson, C.	...Lanowlie
Nicholson, E. F.	...Bombay
O'Connel, J.	...Bombay
Olivier, Captain H. D.	...Bombay
Ommaney, H. T. (c.s.)	...Khandesh
Ormiston, G.	...Bombay
Owen, F. A. ...	...Bombay
Pallis, A. ...	...Europe
Panday, Ardeshir Shapurji	...Bombay
Parker, Capt. J. C.	...Kurrachee
Parmenides, J.	.. Bombay
Patterson, Dr.	...Bombay
Pechey, Miss E. (m.d.)	...Bombay
Penny, Mrs. L.	...Bombay
Peters, Surgeon-Major	...Nasik
Peterson, P. (d.sc.)	...Bombay

NAME.				RESIDENCE.
Peyton, Lieut.-Colonel W.	...	...		...Kanara
Phipson, H. M.	...	...	...	...Bombay
Pinhey, A. F.	...	...	...	...Neemuch
Portman, Major	...	...	...	...Bombay
Powwalla, Sorabjee Cavasjee	...	...		...Bombay
Punnett, F....	...	...	...	...Bombay
Readymoney, N. J...	...	...	...	...Bombay
Reid, G. B. (c.s.)	...	...	...	...Ahmedabad
Ritchie, A. M.	...	...	...	...Beejapoor
Rivett-Carnac, L.	...	...	...	...Bombay
Robb, Surgeon-Major	...	...	...	...Ahmedabad
Robinson, K.	...	...	...	...Allahabad
Russell, L. P.	...	...	...	...Bombay
Salmon, M. B.	...	...	...	...Sadra
Scott, The Hon'ble Justice	...	...		...Bombay
Scott, J. W....	...	...	...	...Bombay
Selby, Captain H. O.	...	...	...	...Bombay
Sheppard, G. F. (c.s.)	...	...	...	...Guzerat
Shipp, W.	...	...	...	...Lanowlie
Simpson, A. F.	...	...	...	...Bombay
Sinclair, W. F. (c.s.)	...	...	...	...Alibag
Slater, E. M.	...	...	...	...Bombay
Sleater, J. M.	...	...	...	...Bombay
Starling, M. H.	...	...	...	...Bombay
St. Clair, Capt. W. A. E. ...	...	...		...Bombay
Sterndale, R. A.	...	...	...	...Bombay
Steward, A. B. (c.s.)	...	...	...	...Poona
Stewart, R. L.	...	...	...	...Bombay
Stuart, C. A.	...	...	...	...Bombay
Sturt, Colonel	...	...	...	...Ahmednugger
Swan, H. H.	...	...	...	...Bombay
Swinhoe, Colonel C.	...	...	...	...Kurrachee
Symons, Mrs. H. S.	...	...	...	...Bombay
Symons, J. L.	...	...	...	...Bombay
Symons, N. S.	...	...	...	...Bombay
Thacker, W.	...	...	...	...Bombay
Thomson, Mrs.	...	...	...	...Bombay
Trevithick, R.	...	...	...	...Bombay
Vidal, G. W. (c.s.)	...	...	...	...Tanna
Walcott, Colonel (c.b)	...	...	...	...Bombay
Wallace, J. ...	...	...	...	...Bombay
Wallace, L. A.	...	...	...	...Bombay
Walton, Rienzi	...	...	...	...Bombay
Ward, Frank	...	...	...	...Bombay
Ward, H. B.	...	...	...	...Bombay
Watson, Rev. A. B.	...	...	...	...Poona

NAME.					RESIDENCE.
Webb, W. ...	...	...	...	...	...Bombay
Weir, Dr. T. S.	...	...	...	...	...Bombay
Wendon, H.	...	...	...	...	...Poona
Wise, H. S....	...	...	...	...	...Bombay
Wright, W.	...	...	...	...	...Bombay
Wroughton, R. C. ...	...	...	...	...	...Poona
Wylie, R. ...	...	...	...	...	...Bhownugger
Yerbury, Major	...	...	...	...	...Punjab
Young, G. S.	...	...	...	...	...Bombay
Young, W. E.	...	...	...	...	...Bombay

Catalogue of the Mammalia in the Collection of the Bombay Natural History Society.

ORDER   I.—QUADRUMANA.—None.
ORDER   II.—LEMURES.—None.
ORDER III.—CARNIVORA.

Family—FELIDÆ.

Genus—FELIS.

Felis tigris (Linn.).—The Tiger. Hab.: Eastern Asia.

    a & b.—Skulls, presented by Mr. J. Shillingford, Purneah.

    c.—A Skeleton, presented by Mr. W. Shipp, Lanowlie.

Felis pardus (Linn.).—The Panther.
Felis domesticus.—Hab. : India.

    a.—Skull. Donor, unknown.

Family—VIVERRIDÆ.

Genus—VIVERRICULA.

Viverricula malaccensis (Gm.).—The Lesser Civet Cat. Hab. : India, Burmah and Ceylon.

    a & b.—Skins prepared for mounting, presented by Mr. E. H. Aitken, Bombay.

Genus—PARADOXUS.

Paradoxurus musanga (F. Cuv.).—The Common Tree-Cat.

    a.—Living specimen from the Straits, presented by Mr. E. Bodger, Bombay.

    b.—Skin and skull, presented by Mr. W. Shipp, Lanowlie.

2

Family—MUSTELIDÆ.
Genus—MARTES.

Martes abietum (Ray).—The Pine Marten. Hab.: Kashmir and Ladak.
a.—Skin, presented by Mr. R. A. Sterndale, Bombay. .

Genus—LUTRA.

Lutra nair (F. Cuv.).—Indian Otter.   Hab.: India.
a.—A skin of an Albino, presented by Mr. W. Shipp, Lanowlie.

ORDER IV.—None.

ORDER V.—CHIROPTERA.

Family—PTEROPODIDÆ.
Genus—PTEROPUS.

Pteropus medius (Temm.).—The Flying Fox.   Hab.: India.
a.—Specimen in spirits, presented by Dr. Charlotte Ellaby, Bombay.

Genus—CYNOPTERUS.

Cynopterus marginatus (Geoffr.).—Little White-eared Fruit Bat. Hab.: India.
a.—In spirits, presented by Mr. E. H. Aitken, Bombay.
b.—Mounted, presented by Mr. J. P. Cornforth, Bombay.

Family—VAMPYRIDÆ.
Genus—MEGADERMA.

Megaderma lyra (Geoffr.).—The Long-eared Vampire Bat. Hab.: India and Ceylon.
a.—Specimen in spirits, presented by Mr. G. W. Vidal, Thana.

Sub-family—RHINOLOPHINÆ.
Genus—HIPPOSIDEROS.

Hipposideros murinus (Elliot).—Little Horse-shoe Bat.   Hab: Southern India, Ceylon and Burmah.
a to d.—In spirits, presented by Mr. E. H. Aitken, Bombay.

Family—NOCTILIONIDÆ.
Genus—TAPHOZOUS.

Taphozous longimanus.—The Long-armed Bat   Hab.: India.
a.—Specimen in spirits, presented by Mr. E. H. Aitken, Bombay.

Family—VESPERTILIONIDÆ.
Genus—SCOTOPHILUS.

Scotophilus Coromandelianus (F. Cuv.).—The Coromandel Bat. Hab: India, Burmah and Ceylon.
a.—Specimen in spirits, presented by Mr. E. H. Aitken, Bombay.

Scotophilus Temminckii (Horsf.).—Temminck's Bat. Hab.: India, Burmah and Ceylon.

a.—In spirits, presented by Mr. E. H. Aitken, Bombay.

b & c.— Do. do. Mr. L. H. Butcher.

Genus—KERIVOULA.

Kerivoula picta (Pall).—The Painted Bat. Hab.: India, Burmah and Ceylon.

a & b.—Two specimens in spirits, presented by Mr. T. Bromley, Junr.

ORDER VI.—RODENTIA.

Family—SCIURIDÆ.

Genus—PTEROMYS.

Pteromys oral (Tickell).—Large Grey Flying Squirrel. Hab.: India and Ceylon.

a to c.—Skins, presented by Col. Peyton.

Pteromys magnificus (Hodgs.).—Hab.: Himalayas. Skin, presented by Mr. J. C. Anderson.

Family—MYOXIDÆ.

Genus—MYOXUS.

Myoxus avellanarius (Linn.).—Common English Dormouse. Hab.: British Islands.

a.—Specimen in spirits, presented by Mr. R. A. Sterndale, Bombay.

Family—MURIDÆ.

Genus—MUS.

Mus decumanus (Pall.).—The Brown Rat. Hab.: Europe and Asia.

a, b & c.—Living specimens, presented by Messrs. E. H. Aitken and R. A. Sterndale, Bombay.

* *Mus sp.?*—Nov. Sp. Hab.: Bombay.

a.—Living specimen, presented by Mr. E. H. Aitken, Bombay.

Mus rattus.—Young, presented by Mr. E. H. Aitken, Bombay.

b & c.—Adult presented by Mr. E. H. Aitken, Bombay.

Mus (Nesokia) Elliotanus.—Elliot's Field Rat. Hab.: Bengal, Assam and Bombay.

a.—Living specimen, presented by Mr. E. H. Aitken, Bombay.

Mus urbanus (Hodgs.)—Common House Mouse. Hab.: India.

a to d.—Living specimens, presented by Mr. R. A. Sterndale, Bombay.

* Apparently undescribed, to be named hereafter.

Family—LEPORIDÆ.

Genus—LEPUS.

Lepus ruficaudatus (Geoffr.).—Common Indian Hare. Hab. : India.

     * *a.*—Specimen mounted in folds of Python. Donor, Mr. W. Shipp, Lanowlie.

ORDER VIII.—PROBOSCIDEA.

Family—ELEPHANTIDÆ.

Genus—ELEPHAS.

Elephasindicus (Cuv.).—Indian Elephant. Hab. : India.

     a.—Skull of male, presented by Mr. J. Shillingford, Purneah.

ORDER IX.—UNGULATA.

Family—RHINOCEROTIDÆ.

Genus—RHINOCEROS.

Rhinoceros indicus (Cuv.).—The Indian Rhinoceros. Hab.: Himalayan Terai, from Central Nepal to Eastern Assam.

     a.—Skull of male, presented by Mr. J. Shillingford, Purneah.

Family—BOVIDÆ.

Genus—GAVÆUS.

Gavæus Gaurus (Ham Smith).—The Gaur or so-called Bison. Hab. : India, in parts.

     **a.*—Mounted head of male, presented by Mr. Leslie Crawford.

     b.—Skull, presented by Mr. Leslie Crawford.

Genus—BUBALUS.

Bubalus Arni (Shaw).—Indian Wild Buffaloe. Hab. : Central India, Terai, from Oude to Bhotan, Assam, Burmah and Ceylon.

     a & b.—Skulls, male and female, presented by Mr. J. Shillingford, Purneah.

Genus—BOSELAPHUS *vel* PORTAX.

Boselaphus tragocamelus (Pall), *Portax pictus* (H. Smith).—The Nylgao or Blue Bull. Hab. : India.

     **a.*—Mounted head, presented by Mr. Leslie Crawford.

Sub-family—ANTELOPINÆ.

Genus—GAZELLA.

Gazella Bennetti (Sykes).—Indian Gazelle. Hab.: India.

     a.—Horns. Donor, unknown.

     b.— Do. presented by Mr. R. A. Sterndale, Bombay.

     c.— Do.   ,,   ,,   ,,

     d & e.—Skins. ,,   ,,   ,,

              * Mounted by Mr. E. L. Barton.

Genus—PANTHOLOPS.

Pantholops Hodgsoni (Abel).—Tibetan Antelope.   Hab. : Tibet.

 a & b.—Mounted heads, purchased.   Dauvergne Collection.

 c to *f.*—Skulls   ,,   ,,

Genus—ANTELOPE.

Antelop ebezoartica (Aldro), *cervicapra* (Pall.).—The Indian Antelope.
Hab. : India.

 a to *c.*—Horns, presented by Mr. R. A. Sterndale, Bombay.

 d.—Mounted head, presented by Mr. J. C. Anderson, Bombay.

 e & f.—Skins, presented by Mr. R. A. Sterndale, Bombay.

Sub-family—RUPI-CAPRINÆ.
Genus—NEMORHÆDUS.

Nemorhœdus bubalina (H. Smith).—The Serow.   Hab. : Himalayas.

 a.—Skull, purchased.   Dauvergne Collection.

Nemorhœdus goral (Hardw.)—The Goral.   Hab. : Himalayas.

 a.—Skull, purchased.   Dauvergne Collection.

Sub-family—CAPRINÆ.
Genus—CAPRA.

Capra megaceros (Hutton).—Markhor.   Hab. : N.-E. Himalayas.

 a & b.—Mounted heads, purchased.   Dauvergne Collection.

 c & d.—Skulls,   ,,   ,,

Capra Sibirica (Meyer).—Himalayan Ibex.   Hab. : Kashmir and
Ladak.

 a to *c.*—Mounted heads, purchased.   Dauvergne Collection.

 d.—Skull,   ,,   ,,

Genus—OVIS.

Ovis Hodgsoni (Blyth).—The Ammon of sportsmen.   Hab. :
Himalayas.

 a.—Mounted head, purchased.   Dauvergne Collection.

 b.—Skull,   ..   ,,

Ovis vignei (Blyth) ⎫
 a.—   ⎪
 b.—   ⎬—The Shapoo. Hab. : N. Himalayas.
 c.—   ⎪ Purchased. Dauvergne Collection.
 d.—   ⎭

Ovis nahura (Hodg.).—The Burhel.   Hab. : N. Himalayas.

 a to *d.*—Heads mounted, purchased.   Dauvergne Collection.

 e & f.—Skulls,   ..   ,,

 g & h.—Skins,

Sub-family—CERVINÆ.

Genus—CERVULUS.

Cervulus muntjac (Temm.), *aureus* (H. Smith).—The Rib-faced Deer or Kakur. Hab.: India.

a.—Mounted head, purchased. Dauvergne Collection.

b.—Skull, presented by Mr. J. Shillingford, Purneah.

Genus—CERVUS.

Cervus cashmirianus (Falc.).—The Kashmir Stag. Hab.: Kashmir.

*a.—Mounted head, purchased. Dauvergne Collection.

b.—   ,.   ,,   ,,

c.—Skull,   ,,   ..

d.—Skin,   ,,   ,,

Cervus (Rucervus) Duvaucelli (Cuv.).—The Swamp Deer. Hab.: Forest lands at foot of Himalayas from the Kyarda Doon to Bhotan, Assam and Central India.

a.—Skull with horns, presented by Mr. J. Shillingford, Purneah.

Cervus (Rusa) Aristotelis (Cuv.).—Sambur. Hab: India.

a.—Skull, presented by Mr. R. A. Sterndale.

b.—Horns. Donor, unknown.

c.—Skull, presented by Mr. J. Shillingford, Purneah.

d.— ,,   ,,   ,,   ,,

Cervus (Axis) Porcinus (Temm.)—The Hog Deer. Hab.: India.

a.—Skull with horns, presented by Mr. J. Shillingford, Purneah.

b.—   ,,   ,,   ..   ,,

Family—TRAGULIDÆ.

Genus—TRAGULUS *vel* MEMINNA.

Tragulus (Meminna) Indica (Erx.).—The Mouse Deer. Hab.: India and Ceylon.

a.—Skin, prepared for mounting, presented by Mr. W. F. Sinclair, Alibag.

Family—SUIDÆ.

Genus—SUS.

Sus indicus (Schinz.) *vel cristatus* (Wagn.)—Indian Boar. Hab·: India, Burmah and Ceylon.

a & b.—Mounted heads, presented by Mr. Barton.

c & d.—Skulls, presented by Mr. J. Shillingford, Purneah.

* Mounted by Mr. E. L. Barton.

ORDER X.—CETACEA.

Family—MYSTICETE.

Genus—BALÆNOPTERA.

Balænoptera India (Blyth).—The Indian Rorqual, or Finback Whale. Hab.: Indian Ocean.

Intervertebral disk. Donor, unknown.

ORDER XII.—EDENTATA.

Family—MANIDÆ.

Genus—MANIS.

Manis Pentadaclyta (Linn.).—The Five-toed Pangolin, or Scaly Anteater. Hab.: India.

a.—Skin, presented by Mr. H. M. Phipson, Bombay.

Catalogue of Birds as yet in the Collection of the Bombay Natural History Society.

(*N.B.*—Contributions in this section are greatly needed, and will be thankfully received.)

An asterisk denotes Mr. Anderson's collection from Simla. A dagger, Colonel W. B. Thomson's, from Cashmere.

ORDER I.—RAPTORES.

Sub-family—FALCONINÆ.

* Cerchneis tinnunculus (*Linn.*).—The Kestrel.

Falco chicquera (*Daud.*).—The Merlin.

Sub-family—ACCIPITRINÆ.

Astur badius (*Gm.*).—The Shikra or Indian Sparrow Hawk.

* Accipiter nisus (*Linn.*).—The European Sparrow Hawk.

Sub-family—AQUILINÆ.

† Pandion halietus (*Linn.*).—The Osprey.

† Haliaetus leucoryphus (*Pall.*).—The Ring-tailed Sea-Eagle.

Sub-family—BUTEONINÆ.

† Buteo ferox (*Gm*).—The Long-legged Buzzard.

Butastur teesa (*Frankl*).—The White-eyed Buzzard.

Sub-family—MILVINÆ.

Haliastur Indus (*Bodd.*).—The Maroon-backed or Brahminy Kite.

Family—STRIGIDÆ.

Sub-family—SYRNIINÆ.

* Syrnium nivicolum (*Hodgs.*).—The Himalayan Wood-Owl.

Sub-family—BUBONINÆ.

Ketupa Ceylonensis (*Gm.*).—Brown Fish Owl.

Sub-family—SURNIINÆ.

* Glaucidium brodii (*Burt.*).—The Collared Pigmy Owlet.

ORDER II.—INSESSORES.

Tribe—FISSIROSTRES.

Family—HIRUNDININÆ.

† Hirundo rustica (*Linn.*).—The Common Swallow.

Sub-family—CYPSELLINÆ.

Collocalia unicolor (*Jerd.*).—Edible Nest Swift.

Family—MEROPIDÆ.

† Merops apiaster (*Linn.*).—The European Bee-eater.
Merops viridis (*Linn.*).—Common Indian Bee-eater.
Merops Philippinus (*Linn.*).—The Blue-tailed Bee-eater.
Merops quinticolor. (*Viell.*).—The Chestnut-headed Bee-eater.

Family—CORACIADÆ.

† Coracias garrula (*Linn.*).—The European Roller.
Coracias Iudica (*Linn.*).—The Indian Roller.

Family—HALCYONIDÆ.

† Alcedo ispida (*Linn.*).—The European Kingfisher.
Alcedo bengalensis (*Gm.*).—Common Indian Kingfisher.
Ceryle rudis (*Linn.*)—The Pied Kingfisher.

Family—BUCEROTIDÆ.

Dichoceros cavatus (*Bodd.*).—The Great Hornbill.

Tribe—SCANSORES.

Family—PSITTACIDÆ.

Sub-family—PALÆORNINÆ.

Palæornis purpureus (*P. L. S. Müll.*).—Rose-headed Parraquet.
* Palæornis schisticeps (Hodgs.)

Family—PICIDÆ.

Sub-family—PICINÆ.

† * Picus himalayensis (*Jerd. and Selb.*).—The Himalayan
                              Pied Woodpecker.
* Picus brunneifrons (*Vig.*).—The Brown-fronted Woodpecker.

Sub-family—CAMPEPHILINÆ.

Chrysocolaptes strictus (*Horsf.*).—Southern Large Golden-backed Woodpecker.

Sub-family—GECININÆ.

* Gecinus squamatus (*Vig.*).—The Scaly-billed Green Woodpecker.

Family—MEGALÆMIDÆ.

Megalæma viridis (*Bodd.*).—Small Green Barbet.

Xantholæma hœmacephala (*P. L. S. Müll.*).—The Crimsonbreasted Barbet.

Family—CUCULIDÆ.

Sub-family—CUCULIDÆ.

† Cuculus canorus (*Linn.*).—The European Cuckoo.

Coccystes jacobinus(*Bodd.*).—The Pied-crested Cuckoo.

Eudynamis honorata(*Linn.*).—The Indian Koel.

Sub-family—PHÆNICOPHAINÆ.

Centrococcyx rufipennis (*Iu.*).—The Common Crow Pheasant.

Tribe—TENUIROSTRES.

Family—NECTARINIDÆ.

Sub-family—NECTARININÆ.

Æthopyga vigorsi (*Sykes*).—The Violet-eared Red Honeysucker.

Cinnyris minima (*Sykes*).—The Tiny Honeysucker.

Cinnyris Asiatica (*Lath.*).—The Purple Honeysucker.

Sub-family—DICÆINÆ.

Dicæum erythrorhynchus (*Lath*).—Tickell's Flower-pecker.

Family—CERTHIADÆ.

Sub-family—CERTHINÆ.

† Certhia Himalayana (*Vig.*).—The Himalayan Tree-creeper.

* Tichodroma muraria (*Linn.*).—The Red-winged Wall-creeper.

Tribe—DENTIROSTRES.

Family—LANIADÆ.

Sub-family—LANIANÆ.

† Lanius erythronotus (*Vig.*).—The Rufous-backed Shrike.

Sub-family—MALACONOTINÆ.

Tephrodornis Pondicerianus (*Gm.*).—The Common Woodshrike.

Sub-family—CAMPEPHAGINÆ.

Grauculus macii (*Less.*).—The Large Cuckoo-shrike.

Pericrocotus flammeus (*Forst.*).—The Orange Minivet.

Pericrocotus peregrinus (*Linn.*).—The Small Minivet.

* Pericrocotus brevirostris (*Vig.*)—The Short-billed Minivet.

Sub-family—DICRURINÆ.

Buchanga atra (*Herm.*).—The Common Drongo-shrike or King-crow.

Buchanga cærulescens (*Linn.*).—The White-bellied Drongo.

Family—MUSCICAPIDÆ.

Sub-family—MYIAGRINÆ.

† Muscipeta paradisi (*Linn.*).—The Paradise Fly-catcher.

Sub-family—MUSCICAPINÆ.

Cyornis tickelli (*Blyth.*).—Tickell's Blue Redbreast.

† Muscicapula Superciliaris (*Jerd.*).—The White-browed Blue Fly-catcher.

Family—MERULIDÆ.

Sub-family—MYISTHERINÆ.

Myiophoneus Horsfieldi (*Vig.*).—The Malabar Whistling Thrush.

† Myiophoneus Temminckii (*Vig.*).—The Yellow-billed Whistling Thrush.

Sub-family—MERULINÆ.

* Petrophila erythrogastra (*Vig.*).—The Chestnut-billed Thrush.

Petrophila cinclorhyncha (*Vig.*).—The Blue-headed Chat Thrush.

† Cyanocinclus cyanus (*Linn.*).—The Blue Rock Thrush.

Geocichla cyanotis (*Jerd. & Selb.*).—The White-winged Ground Thrush.

† Geocichla unicolor (*Tickell*)—The Dusky Ground Thrush.

Merula nigropilea (*Lafr.*)—The Black-capped Blackbird.

* Turdus ruficollis (*Pall.*).—The Red-tailed Thrush.

* Oreocincla mollissima (*Bly.*)—The Plain-backed Mountain Thrush.

Sub-family—TIMALINÆ.

Pomatorhinus Horsfieldi (*Sykes*).—The Southern Scimitar Babbler.

* Trochalopterum variegatum (*Vig.*).—The Variegated Laughing Thrush.

* Trochalopterum lineatum (*Vig.*).—The Streaked Laughing Thrush.

Malacocercus Somervillii (*Sykes*).—The Rufous-tailed Babbler.
* Malacias capistratus (*Vig.*).—The Black-headed Sibia.
Chatarrhœa candata (*Dum.*).—The Striated Bush Babbler.

Family—BRACHYPODIDÆ.

Sub-family—PYCNONOINÆ.

Otocompsa fuscicaudata (*Jerd.*).—The Southern Red-whiskered Bulbul.
* † Otocompsa leucogenys (*Gray*).—The White-checked Crested Bulbul.
Molpastes hæmorrhous (*Gm.*).—The Common Madras Bulbul.
* † Hypsipetes psaroides (*Vig.*).—The Himalayan Black Bulbul.

Sub-family—PHYLLORNITHINÆ.

Phyllornis Jerdoni (*Blyth*).—The Common Green Bulbul.
Phyllornis Malabaricus (*Gm.*).—The Malabar Green Bulbul.
Iora tiphia (*Linn.*).—The Black-headed Green Bulbul.

Sub-family—ORIOLINÆ.

Oriolus Galbula (*Linn.*).—The Golden Oriole.
Oriolus melanocephalus (*Linn.*).— The Bengal Black-headed Oriole.
† Oriolus Kundoo (*Sykes*).—The Indian Oriole.

Family—SYLVIADÆ.

Sub-family—SAXICOLINÆ.

Thamnobia cambaiensis (*Lath.*).—The Brown-backed, or Southern India Robin.
Pratincola caprata (*Linn.*).—The White-winged Bush Chat.
* Pratincola macrorynchus (*Stol.*).—The Long-billed Bush Chat.

Sub-family—RUTICILINÆ.

Chæmorrornis lencocephalus (*Vig.*).—The White-capped Redstart.

Sub-family—DRYMOICINÆ.

Drymœca inornata (*Sykes*).—The Earth-brown Wren Warbler.
Drymœca insignis (*Hume*)·—The Great Wren Warbler.
Franklinia Buchanani (*Blyth*).—The Rufous-fronted Wren Warbler.

Sub-family—PHYLLOSCOPINÆ.

† Reguloides occipitalis (*Jerd.*).—The Large Crowned Warbler.
* Reguloides proregulus (*Pall.*).—The Crowned Tree Warbler.
* Abrornis albo-superciliaris (*Bly.*).—The White-browed Warbler.

Sub-family—SYLVINÆ.

† Sylvia affinis (*Bly.*).—The Allied Grey Warbler.

Sub-family—MOTACILLINÆ.

† Budytes calcaratus (*Pall*).—The Yellow-headed Wagtail.

† Agrodroma sordida (*Rüpp*).—The Brown Rock Pipit.

Sub-family—LEIOTRICHINÆ.

* Siva strigula*(*Hodgs.*).—The Stripe-throated Hill Tit.

* Minla castaniceps (*Hodgs.*).—The Chesnut-headed Hill Tit.

Sub-family—PARINÆ.

* Lophophanes melanolophos (*Vig.*).—The Crested Black Tit.

* Parus monticolus (*Vig.*).—The Green-backed Tit.

† Parus nipalensis (*Hodgs.*).—The Indian Grey Tit.

Tribe—CONIROSTRES.

Family—CORVIDÆ.

Sub-family—CORVINÆ.

* Corvus macrorhyncus (*Wagler*).—The Indian Corby, or Carrion
Crow.

Corvus splendens (*Vieill.*).—The Common Indian Crow.

* Nucifraga hemispila (*Vig.*).—The Himalayan Nut-cracker.

† Nucifraga multipunctata (*Gould.*).—The Spotted Nut-cracker.

Sub-family—GARRULINÆ.

† Pica bottanensis (*Deless*).—The Himalayan Magpie.

* Garrulus bispecularis (*Vig.*).—The Himalayan Jay.

† Urocissa flavirostris (*Bly.*).—The Yellow-billed Blue Magpie.

Sub-family—DENDRCOITTINÆ.

Dendrocitta rufa (*Lath.*).—The Common Indian Magpie.

† Dendrocitta Himalayensis (*Bly.*).—The Himalayan Magpie.

Graculus eremita (*Linn.*).—The Himalayan Chough.

Family—STURNIDÆ.

Sub-family—STURNINÆ.

† Sturnus nitens (*Hume*).—The Glossy Black Starling.

Acridotheres tristis (*Linn.*).—The Common Myna.

Acridotheres fuscus (*Wagler*).—The Dusky Myna.

Sturnia pagodarum (*Gmel.*).—The Black-headed Myna.

Sub-family—FRINGILLINÆ.

† Carpodacus erythrinus (*Pall.*).—The Common Rose Finch.

* Pycnoramphus icterioides (*Vig.*).—The Black and Yellow Grosbeak.

|Sub-family—ESTRELDINÆ.

Amadina rubronigra (*Hodgs.*).—The Chestnut-bellied Munia.

Sub-family—ALAUDINÆ.

Pyrrhulauda grisea (*Scop.*).—The Black-bellied Finch Lark.

Alauda gulgula (*Frankl.*).—The Indian Sky Lark.

Sub-family—PASSERINÆ.

† Passer domesticus (*Linn.*).—The Common Sparrow.

* Passer cinnamomeus (*Gould*).—The Cinnamon-headed Sparrow.

Sub-family—EMBERIZINÆ.

† * Emberiza stracheyi (*Moore*).—The White-necked Bunting.

† Emberiza Stewarti (*Bly.*).—The White-capped Bunting.

† Emberiza fucata (*Pall.*).—The Grey-headed Bunting.

ORDER III.—GEMITORES.

Family—TRERONIDÆ.

Sub-family—TRERONINÆ.

Osmotreron Malabarica (*Jerd.*).—The Grey-fronted Green Pigeon.

Sub-family—TURTURINÆ.

Turtur Suratensis (*Gm.*).—The Spotted Dove.

ORDER IV.—RASORES.

Family—PTEROCLIDÆ.

Pterocles exustus (*Tem.*).—The Common Sandgrouse.

Family—PHASIANIDÆ.

*Pucrasia macrolopha (*Less.*).—The Pakras Pheasant.

† Pucrasia castanea (*Gould*).—

* Euplocomus albocristatus (*Vig.*).—The White-crested Kalij Pheasant.

Sub-family—GALLINÆ.

Galloperdix spadiceus (*Gm.*).—The Red Spur Fowl.

Family—TETRAONIDÆ.

Sub-family—PERDICINÆ.

Francolinus pictus (*Jerd and Selb.*).—The Painted Partridge.

* Caccabis Chukor (*Gray*).—The Chukor Partridge.

* Arboricola torqueola (*Val.*).—The Black-throated Hill Partridge.

Sub-family—COTURNICINÆ.

Coturnix Coromandelica (*Gm.*).—The Black-breasted, or Rain Quail.

ORDER V.—GRALLATORES.

Tribe—PRESSIROSTRES.

Family—OTIDIDÆ.

Sypheotides aurita (*Lath.*).—The Lesser Florican.

Family—CHARADRIDÆ.

Sub-family—VANELLINÆ.

Lobivanellus Indicus (*Bodd.*).—The Red Wattled Lapwing.

Tribe—LONGIROSTRES.

Family—SCOLOPACIDÆ.

Sub-family—TRINGINÆ.

Tringa Temminckii (*Liesl.*).—The White-tailed Stint.

Sub-family—TOTANINÆ.

Rhyacophila glareola (*Linn.*).—The Spotted Sandpiper.

Family—PARRIDÆ.

Sub-family—PARRINÆ.

† Hydrophasianus chirurgus (*Scop.*).—The Pheasant-tailed Jacana.

Tribe—CULTIROSTRES.

Family—ARDEIDÆ.

Bubulcus Coromandus (*Bodd.*).—The Cattle Egret.

Ardeola grayi (*Sykes*).—The Pond Heron.

Demi-egretta gularis (*Bosc.*).—The Ashy Egret.

ORDER—NATATORES.

Tribe—LAMELLITOSTRES.

Family—ANSERIDÆ.

Sub-family—PLECTROPTERINÆ.

Sarcidiornis melanonotus(*Penn.*).—The Nukta or Black-backedGoose.

Tribe—MERGITORES.

Family—PODICIPIDÆ.

† Podiceps minor (*Gm.*).—The Little Grebe, or Dabchick.

Tribe—VAGATORES.

Family—LARIDÆ.

Sub-family—STERNINÆ.

Sterna Seena (*Sykes*).—The Large River Tern.

† Sterna melanogastra (*Temm.*).—The Black-bellied Tern.

The following Eggs were received chiefly from Mr. Davidson:—

Gyps pallescens.	Syrnium ocellatum.
Neophron ginginianns.	Bubo bengalensis.
Falco jugger.	Carine brama.
Astur badius.	Hirundo filifera.
Aquila vindhiana.	Hirundo erythropygia.
Nicaetus fasciatus.	Hirundo fluvicola.
Limnaetus cirrhatus.	Collocalia unicolor.
Butastur teesa.	Dendrochelidon coronata.
Haliastur indus.	Ptyonoprogne concolar.
Milvus govinda.	Cypsellus affinis.

Caprimulgus asiaticus.
Caprimulgus monticolus.
Merops viridis.
Merops philippinus.
Coracias indica.
Halcyon smyrnensis.
Alcedo bengalensis.
Ocyceros birostris.
Picus mahrattensis.
Yungipicus nanus.
Brachypternus aurantius.
Megalaima inornata.
Coccyestes jacobinus.
Eudynamis honorata.
Centrophus rufipennis.
Cinnyris asiatica.
Lanius lahtora.
Lanius erythronotus.
Lanius vittatus.
Tephrodornis pondicerianus.
Volvocivora sykesi.
Pericrocotus peregrinus.
Pericrocotus erythropygius.
Buchanga atra.
Leucocerca leucogaster.
Myiophoneus horsfieldi.
Pcytoris sinensis.
Malacocerus terricolor.
Argya malcolmi.
Chatarrhæa caudata.
Ixus luteolus.
Otocompsa fuscicaudata.
Molpastes hæmorrhaus.
Iora tiphia.
Oriolus kundoo.
Thamnobia fulicata.
Thamnobia cambaiensis.
Orthotomus sutorius.
Prinia stewarti.
Prinia hodgsoni.

Drymœca inornata.
Drymœca rufescens.
Fraulinia buchanani.
Motacilla Maderaspatna.
Corvus macrorhynchus.
Zosterops palpebrosa.
Corvus splendens.
Dendrocitta rufa.
Acridotheres tristis.
Acridotheres ginginianus.
Sturnia pagodarum.
Ploceus philippinus.
Amadina punctulata.
Amadina malabarica.
Estrelda amandava.
Gymnoris flavicollis.
Mirafra erythroptera.
Pyrrhulauda grisea.
Spizalauda deva.
Pterocles exustus.
Galloperdix spadiceus.
Francolinus pictus.
Ortygornis ponticerianus.
Perdicula asiatica.
Coturnix coromandelica.
Turnix tiagoor.
Lobivannellus indicus.
Lobipluvia malabarica.
Œdicnemus scolopax.
Parra indica.
Erythra phœnicura.
Hypotænidia striata.
Herodias garzetta.
Ardeola grayi.
Ardea cinnamomea.
Tantalus leucocephalus.
Sarcidiornis melanonotus.
Podiceps minor.
Pelecanus philippensis.

Note by the Editors.—WE have so far catalogued our Mammals and Birds subject to additions in the future which will be noted from time to time. As yet we have not been able to complete our lists of Fishes, Reptiles, &c., which have been reserved for our next issue, but we may briefly state that our collections up to date consist of—

   257 Specimens, comprising about 200 species of Fish in spirits.
    12 Fishes stuffed and mounted by Mr. H. M. Phipson.
    83 Specimens of Snakes in spirit.
    47 Other Reptiles in spirit.
    42 Crabs in spirit.
    53 Crabs dried and set.
    71 Other Marine Animals in spirits.

In addition to the above we have a collection of Butterflies from the Bombay Presidency, the Malabar Coast, the Himalayas, the Punjab and from Aden; also some Moths, Beetles and other insects at present undergoing classification.

THE SOCIETY'S LIBRARY

Contains as yet but the following books :—

MAMMALS.

Mammals of India—(Jerdon.)
Mammalia of India and Ceylon—(Sterndale.)
Histoire Naturelle de Mammiferes—(Gervais.)

BIRDS.

Birds of India—(Jerdon.)   3 Vols., 2 copies.
Stray Feathers—(Hume, ed.).   7 Vols.
Birds' Nesting in India—(Marshall.)
Birds of British Burmah—(Oates.)
Fauna Japonica, Aves—(Siebold.)
Birds of South Africa—(Layard & Sharpe.)
Monograph of the Sunbirds—(Shelley.)
Monograph of the Birds of Paradise—(Elliott.)
Monograph of the Jacamars—(Sclater.)

REPTILES AND FISHES.

Reptiles of India—(Günther.)
Indian Snakes—(Nicholson.)
Malabar Fishes—(Day.)   2 Copies.
Fresh Water Fishes of India—(Beaven.)
Fishes of Madeira—(Lowe.)
Fauna Japonica, Reptilia et Pisces—(Siebold.)
Fishes of the Coromandel Coast—(Russell.)

INSECTS.

Classification of Insects—(Westwood.)
Text Book of Entomology—(Kirby.)
Butterflies of Great Britain (Westwood.)
The Aurelian—(Harris.)
Encyclopédie d'Histoire Naturelle, Papillons—(Cherin.)

OTHER INVERTEBRATA.

Conchology—(Lammarck.)
Fauna Japonica, Crustacea—(Siebold.)

BOTANY.

Flora of British India—(Hooker.)
Ferns of British India—(Beddome.)
Vegetable Products of the Bombay Presidency—(Birdwood.)
Bombay Flora—(Dalzell & Gibson.)

Plants and Drugs of Sind—(Murray.)
Timber Trees of India—(Balfour.)
Flore Forestière de Cochin Chine—(Pierre.) 5 parts.
Icones Plantarum—(Wight.). Vols. II. to VI.

GENERAL.

Zoological Atlas—(Brehm.)
Museum of Natural History.
Vertebrata of Sind—(Murray.)
Cassels' Natural History.
Naturalist's Wanderings in the China Seas—(Collingwood.)
The Calcutta Journal of Natural History. 7 Vols.
Beeton's Dictionary of Natural History.
Naturalist's Wanderings in the Eastern Archipelago—(Forbes.)
Anatomy of Vertebrated Animals—(Huxley.)
The *Asian*—Vols. I. to VI.
Odontography—(Owen.)
Lectures on Comparative Anatomy—(Owen.)

NOTE ON AN UNDESCRIBED HAMALOPSIDA,

By the Rev. F. Dreckmann, s.j.

Figured on stone by Mr. R. A. Sterndale.

This specimen was forwarded to the Society from Saugor, Central Provinces, by Mr. H. Craufuird Thomson.

Head short, thick, broad, distinct from neck; cleft of mouth turned upwards behind; eyes small with round pupil; nostrils on the upper surface of the head in a single large nasal shield, the outer part of which is divided by a groove running outwards from the nostrils; the two nasals contiguous; two small anterior frontals; two loreals, one above the other, the lower one larger than anterior frontals, the upper one small, vertical, longer than broad, five-sided—one præocular, two postoculars; 8 upper labials rather high, the 4th entering the orbit; temporals, $1 + 2 + 0.0$; anterior chin shields in contact with four lower labials; posterior chin shields small, scale-like; six transverse series of scales between chin shields and first ventral. Scales smooth, 33; ventrals narrow, 158; anals and sub-caudals bifid. Sub-caudals, 54; ground colour yellowish white, with 32 large irregular rounded black spots, leaving a narrow stripe of ground colour between them;

4

a series of rather irregular black spots along lower part of the side, alternating with vertebral spots, so that the ground colour appears as decussating stripes. Belly densely checkered with black. Two yellowish lines on the upper part of the head diverging from the muzzle over the eyes to the sides of the head ; from each side of the vertical a line diverging towards the occiput. Length 10 inches, of which the tail is 1½ inches.

NOTE ON A PROBABLE NEW SPECIES OF IBEX.

(*Capra Dauvergnii, nob.*) or variety of *Capra sibirica*.

By R. A. Sterndale, f.z.s.

I have always been averse to multiplication of species, and the tendency of modern research has been to diminish the number of existing sub-divisions; it is, therefore, with some hesitation I bring forward the claims of the subject of my note to separation from the two known species of Oriental Ibex—*Capra sibirica* and *Capra Ægagrus* ; *Capra skeen* and *Capra himalayica* of authors being identical with *C. sibirica*. The question of hybridization between the various marked species of Capræ and Oves has not as yet received the attention that it should, and I think on examination it will be found that certain named species will prove to be hybrids, notably *Ovis Brookei*, but on this point I shall have more to say on a future occasion. The horns of which I have given an illustration herewith were purchased for me some months ago in Kashmir by my friend Mons. H. Dauvergne, simply on account of their size, being 52 inches in length. On receiving them, I was struck by their remarkable divergence from the types of *C. sibirica* and *Ægagrus* and from any Ibex horn I had ever seen. At first I took it to be a hybrid between the two abovementioned species, but I subsequently abandoned this idea, for it bears no resemblance at all to the latter beyond the departure from the usual curve of the well-known Himalayan Ibex. Eccentric forms are not uncommon in the Persian animal, and Mr. Danford figures a pair, in his article in the P. Z. S. for 1875, page 458, the tips of which turning inwards cross each other. In my specimen the horns sweep backwards and outward having widely divergent tips, and in a case of hybridization I should look for some modification of the section of the horn, the two species being so vastly different in this respect. *Capra sibirica* has a square horn, the front broader than the back, and strongly marked with transverse ridges at intervals of less than an inch. *C. œgagrus* has an oval or compressed elliptic section, flatter on the inner side and with a sharp keel or longitudinal ridge in front, which ridge has irregular knobs at considerable distances. In the horns under notice the section resembles that of *sibirica*, flat in front and at the sides, slightly rounded beneath. In texture and color resembling the Markhor horn, being much darker than the ordinary Ibex. Instead of the decided ridges of *C. sibirica* there are rugosities or folds at the following distances:—

Right horn.........$2\frac{4}{16}'$, $2\frac{4}{16}'$, $3\frac{4}{16}'$, $4'$, $5\frac{1}{16}'$, $5\frac{4}{16}'$.

Left horn ....$2\frac{3}{16}$, $2\frac{1}{16}$, $2\frac{4}{16}$, $4\frac{1}{16}$, $5\frac{11}{16}$, $5\frac{4}{16}$.

the rest of the horn is more decidedly and closely knobbed, with fine rings at the tips.

At the base the section measures about $3\frac{1}{2}''$ from front to back, and about $2\frac{3}{4}''$ across.

As regards the skull, which in my specimen is damaged in the occipital region, it is somewhat slighter than that of a head of *C. sibirica* with 40 inch horns, but until we get more specimens to work upon, it is useless to dwell on skull characteristics in this paper. Colonel Kinloch, in a letter to the *Asian*, seems to think that this is an abnormality or sport, from an abnormal specimen killed by him, but from what I gather I am inclined to believe that my skull is of a distinct species or variety from the hills north of the Kishengunga river. Mons. Dauvergne wrote to me in August last that he came across a similar head two years ago, freshly killed, horns measuring 42 inches, widely divergent like mine, with the same characteristics of smoothness and section. I hope to see this head some day as it has been traced. In a recent letter, dated 13th December, he says: "Another horn of the same tribe has been purchased by Sir Oliver St. John; that is the third I know of, and those skin men tell me they have seen them often, but that they are not numerous." Their habitat is as yet uncertain, but with such an enthusiastic and experienced sportsman as Mons. Dauvergne on the scent, I do not despair of deciding this question. He thinks they may come from the range of hills north of the Kishengunga river or the Khagan county west of Kashmir. From enquiries I have made, there are no similar horns in the British Museum or in the

India Museum at Calcutta. I have also received letters from Sir Victor Brooke and Mr. W. T. Blanford, both high authorities on Indian ruminants, to the effect that such a formation is new to them. Mr. Blanford suggests that it might be a hybrid between *C. sibirica* and *megaceros*, variety *Falconeri;* but, though I have gone into the records of hybridization in the gardens of the Zoological Society, and therefore believe such a combination possible, the absence of any flattening of the horn and also its curvature is against the theory. The discovery of more than the three heads above mentioned and the fixing of a particular locality will go far towards proving the existence of a distinct species. In the meantime, in placing this head on record, I wish to associate with it the name of Mons. Dauvergne, to whom both the Society and myself are under considerable obligations as regards assistance in obtaining specimens of Kashmir fauna, and who first brought it to my notice, and therefore, as a tentative measure, I propose to call it that of "*Capra Dauvergnii.*"—R. A. S.

NOTE ON MYGALE FASCIATA.

By Capt. T. R. M. Macpherson.

2nd February.—The following extract was read from a letter which the Secretary had received from Captain T. R. M. Macpherson, forwarding ten specimens of a very large species of *Mygale* found by him in the Kamora district:—

"The spiders, though fairly common in the evergreen forests of this district, are little known, and few men have ever seen them. The first I ever saw I found in one of my boots last year, and shortly afterwards I discovered their habits. They are, I think, entirely nocturnal, keeping always to their burrows in the day time. I have not been able to ascertain what they live on, but it is probably lizards and small birds. The natives call them *Wagh Duri,* and say that they are very venomous, but I have never heard of anybody being bitten by them. However, their long, sharp and hollow 'falces,' strongly resembling the poison fangs of a snake, lead me to believe that they are poisonous, and I would recommend caution in handling them. I experimented the other day on a chicken. The spider attacked it viciously, drawing blood in several places, but beyond frightening the chicken there was no result. However, this spider may have exhausted its poison, for it had been much irritated and had been striking repeatedly at sticks and other things before it attacked the chicken.

"These spiders live in burrows which they excavate in steep banks of earth. The burrows vary from one inch to 2½ inches in diameter, in accordance with the size of the occupant, and are of the form shown below." (The diagram showed a short, straight passage, turning sharply to one side at the end, which was a little widened to form a chamber.) "Sometimes the chamber is to the right, instead of to the left, as shown in the diagram. The burrow and chamber are lined throughout with a closely woven, soft web, much resembling very fine white tissue paper. The total length of the burrows averages, I should say, about 15 inches, the straightpart being about a foot, and excavated perpendicularly to the face of the bank, which is usually precipitous."

Editor's Note.—This species which appears to be *Mygale fasciata* (Seba) is not uncommon in Southern India and Ceylon, but has not attracted much attention from its nocturnal habits. There has been a controversy of long standing regarding the bird-eating propensities of this genus. The first to give currency to the assertion was Madame Merian, who, in a work on Surinam Insects, published in 1705, figured *Mygale avicularia* in the act of devouring a bird. In 1834 Mr. Macleay (P. Z. S. for that year, page 12) threw doubt on her accuracy, and disbelieved in any bird-catching spider, which opinion, however, he subsequently modified (*Ann. and Mag. Nat. Hist.,* 1842, Vol. VIII., p. 324), having seen in Australia a large *Epeira diadema* sucking the juice of a small bird, *Zosterops dorsalis,* which it had caught in its net, but he was still inclined to think it exceptional and accidental. However, other writers have since supported Madame Merian. (See same vol. *Ann. and Mag. Nat. Hist.,* p. 436.) Mons. Jonnes says that its mode of attack is to throw itself on to its victim, clinging by the double hooks of the tarsi, and striving to reach the back of the head to insert its jaws between the skull and the vertebræ. Sir Emerson Tennent was told by a lady who lived near Colombo that she had seen a *Mygale* devour a house lizard. Mr. Edgar Layard (*Ann. and Mag. Nat. Hist.,* May, 1853,) described a fight between a *Mygale* and a cockroach, not much of a fight, for the poor cockroach was speedily overcome and devoured. Mr. Bates, the author of the "Naturalist on the Amazons," has stated that he has seen birds entangled in webs spun by a species of *Mygale* and the spider actually on the bird, and his opinion was that if the *Mygales* did not prey upon vertebrated animals he could not see how they could find sufficient subsistence. (*The Zoologist,* Vol. XIII., p. 480.) So far evidence is in favour of the bird-eating propensities of this genus, but it would be interesting to prove the habits of our Indian species, and therefore living specimens taken, if possible with nest complete, would be most acceptable in order that they may be placed under observation.—R. A. S.

ON THE MIMICRY SHOWN BY PHYLLORNIS JERDONI.

By Mr. E. H. Aitken.

At the Meeting of the 1st Section held on July 30, 1885, Mr. E. H. Aitken put in the following note:—

"On two occasions lately my attention has been attracted to the extraordinary powers of mimicry possessed by the green Bulbul, *Phyllornis Jerdoni* or *Malabaricus*, I am uncertain which, as both are found on the Western Ghâts, and I did not in either case see the bird clearly enough to distinguish it. In May I was walking up from Narel to Matheran when I heard the notes of several familiar birds in one bush. I threw stones into the bush and a pair of green Bulbuls flew out. There was nothing else in the bush. On the second occasion, last July, I was at Tanna seeking for nests, when I thought I heard *Malacocercus Somervillei* in a jambool tree. I went up to the tree and could see no bird, but the *Malacocercus* continued very noisy. Then I heard a King Crow, *Buchanga Atra*, calling out vigorously. I pelted the tree with stones, and after a little a green Bulbul appeared at the very top of the tree and began to abuse me in several languages. Jerdon quotes Tickell to the effect that P. Jerdoni is an excellent mocking bird, but as he does not support the statement by his own experience, and as no other writer I know of mentions the fact, I think it is worthy of notice."

JOURNAL

OF THE

BOMBAY

𝔑atural 𝔥istory 𝔖ociety.

| No. 2. | BOMBAY, APRIL 1886. | Vol. I. |

NOTES ON "THE BIRDS OF BOMBAY," BY LIEUT. H. E. BARNES.*

(BY H. LITTLEDALE, Baroda.)

IT would be presumptuous for a mere tyro in ornithology like myself to attempt a detailed or formal criticism of Mr. Barnes's book. It seems to me, speaking generally, to be very fairly done, and to furnish, what many sportsmen and naturalists will be glad to have, a cheap and comprehensive descriptive catalogue of the birds of the Presidency. Until its publication, naturalists have had to get "Jerdon's Birds," "Stray Feathers," Hume and Marshall's "Game-birds," Sharpe's "Catalogue of Birds," Hume's "Nests and Eggs," and other books, costing in all about Rs. 400, and requiring a book-case to hold them, besides entailing much labour to search out particular birds. Now we have this handy and well-printed volume, that will give most that we want for ordinary purposes, will go into a game bag, and costs only Rs. 8.

Besides presenting the descriptions and measurements of birds as found in Jerdon and other writers, this Handbook contains the results of Mr. Barnes's twenty years' work at the birds of this Presidency, and I have read these scattered observations with so much interest that, like Oliver Twist, I cannot help asking for more. And that "more" Mr. Barnes might certainly have given me if he had called on his fellow-ornithologists of this Society to let him have the use of their field note-books for his work; in records of, and deductions from, field ornithology, the more workers the better work. Hence, if, in the following hastily jotted remarks, I appear to grumble some-what, it is in no fault-finding or ungrateful spirit; my object is addition, not subtraction, and I heartily thank Mr. Barnes for what he has done, and advise every good sportsman to buy and study his book.

---

* *Handbook to the Birds of the Bombay Presidency*, by Lieut. H. EDWIN BARNES, D.A.C., Central Press, Calcutta, 1885. Price Rs. 8-8, V.P.P.

Passing over questions of nomenclature, and shunning such a
Charybdis as the discrimination of difficult species (like Aquila
Nævia), I consider that, in the remarks on distribution and on nests
and eggs in particular, Mr. Barnes has not only lost much good
material that our birds' nesting members would gladly have contributed, but he has also not made as good use of his actual authorities as
he might have done.   The care and fulness with which the nidification
of many birds is described, make me wonder that nothing is said of the
nests and eggs of many other birds, which are all more or less fully
dealt with in books that Mr. Barnes had before him when compiling
his work. For instance, Mr. Barnes says that he has been unable to ascertain anything about the breeding of *Elanus cæruleus*, the black-
winged kite, whereas there is a full account (from the competent pen of
Mr. Davidson) in "Stray Feathers," Vol. viii., pp. 370 and 415,
to say nothing of Bree's "Birds of Europe," Vol. i.

I may add that, on the 23rd October 1885, I had a nest with three
hard set eggs taken, and the two birds shot, at Tandalja, two miles
from Baroda, and that earlier in the same month I found a nest at
Tatarpura, six miles from Baroda, with young birds in it. The eggs
were, as Mr. Hume says somewhere, like "miniature Neophrons,"
and not like Dr. Bree's figures.

Again, while describing the eggs of the *Prinias*, Mr. Barnes omits to
point out that the mahogany-coloured eggs are laid by the species with
ten tail-feathers, while the birds with *twelve* tail-feathers lay eggs of a
different type. Of the eggs of *Prinia gracilis* and *P. Hodgsoni* (the
two species are, I am convinced, identical, the latter being the
breeding plumage), Mr. Barnes says not a word. The eggs are
remarkable, being of two types of ground colour, *viz.*, pure white
and pale blue, and being either unspotted, or speckled with light
red. There are, therefore, four varieties of the eggs of these tiny
birds. They are very common about Baroda, and breed along the
railway line. The eggs in all the twenty odd nests I found last
August were uniform in each nest, *i.e.*, all in each nest were either
pure white, or pure bluish, or white, speckled red, or blue, speckled
red, but I have found the several types in different nests only a few
yards apart, and could see no external difference in the birds.
Again, Mr. Barnes is rather careless in saying of the tailor-bird,
O. sutorius, that "occasionally the eggs are of a greenish white
colour." There are (as Mr. Hume has pointed out) two types of
ground colour, either pure white or pale greenish blue, but *both* types

are blotched with red-brown. Mr. Barnes's words would lead one to suppose that the latter type was occasionally without markings, which is never the case. His description of the tailor-bird, I may add, does not discriminate the sexes sufficiently.

With regard to the nesting of the common Indian Swift (*C. affinis*) I may add to Mr. Barnes's observations the curious fact that on the 23rd February 1885 the nests of a colony of the cliff swallow (*H. fluvicola*) under the City Bridge, Baroda, were found by me to be·occupied by about fifty of these swifts, who had eggs and youngs in them, while the cliff swallows had been forced to build a fresh cluster of nests further under the arch for their February brood. The nests the swifts had taken were probably those built by the cliff swallows for their previous September clutch, as last October I found that the young cliff swallows had all just flown and that a few young swifts were still unfledged in the nests of *H. fluvicola*.[*]

Speaking of the swifts, Mr. Barnes calls *C. melba* (the Alpine Swift), a somewhat rare cold weather visitant. I saw seven and shot one near Baroda on the 21st September 1885, which is well before the "cold weather."

Since the publication of Captain Marshall's useful book, "Bird's Nesting in India," in which the eggs of *Caprimulgus Mahrattensis* are stated to be unknown, Mr. Doig found them to be common in Sind, and described them in "Stray Feathers" (Vol. viii., p. 372). Mr. Barnes does not describe the eggs, which, out of Sind at least, would be a valuable find for an oologist.

The blue-tailed bee-eater, says Mr. Barnes, "occurs sparingly throughout our district." It is common along the Guzerat rivers, and I have seen hundreds along the Mahi from Wasad to Dabka. They move to the tanks and meadows, especially those near the tele-graph wires, in the rains, returning to the larger rivers as the country dries up. I took thirty eggs last year from deep holes in nullahs along the Mahi—eggs like those of *M. viridis*, but larger. In Guzerat the common bee-eater is called *tilwa* : Mr. Barnes gives "hurrial" as the Hind. name. I do not see what Hind. names have to do with the Bombay Presidency. A guide to the birds of this side of India should give the names in as many as possible of the local vernaculars, and should be rich in such details. Mr. Barnes's book is very deficient in this respect, and I would suggest that our Society

---

[*] Canon Tristram *Fauna and Flora of Palestine*, p. 84, notes the same of *C. affinis*. I may remark "that the Baroda swifts had not made any addition|of an aggluti-nated straw and feather entrance to the original edifice of clay," as in Palestine.

might compile a list of the Marathi, Guzerati, Sindhi, Canarese, Bhil, &c., names of the better-known birds. I have already made a beginning at such a list of the Guzerati names.

The Indian stork-billed kingfisher, *P. gurial*, has not, Mr. Barnes says, been recorded from Guzerat. Certainly it is not in Captain. Butler's list, but I shot one in a banyan tree on the bank of Jaoli tank, 20 miles north of Baroda, on the 3rd of November last, and Mr. Davidson writes to me that "this species breeds at Godhra behind the Collector's bungalow." The little Indian kingfisher, Mr. Barnes says, lays five or six eggs. Last year I three times found seven eggs in a nest. On the 27th August 1884, in the middle of the rains, I found a nest with five fresh eggs near my house : about three months later than they are usually supposed to breed. Mr. Barnes is partly mistaken in saying that the Pied Kingfisher *never* resorts to wells or tanks. On the tanks hereabouts they reside and breed commonly. And why does Mr. Barnes tell us nothing about the wonderful breeding habits of the Hornbills ?

As regards the koel, every naturalist has a different tale to tell ; but I have found koel's eggs in crow's nests in which there was no crow's egg: it seems improbable that the koel would have laid in an empty nest. Once I actually found near Baroda four koel's eggs, ready to hatch, in a crow's nest in which there was no crow's egg ! This looks as if the koel, sometimes at least, removed the crow's eggs, unless, indeed, we suppose that the crow having no family of her own had adopted the koel's ! Birds do such queer things ! I once found a Pariah kite sitting close on a hare's skull !

On page 137, No. 235 is misprinted 205, and I remember noticing an unnecessary *d* in the middle of *Blanford* somewhere. No. 238, *Dicœum minimum*, I have several times met with here, and I have found one nest, which was, however, deserted afterwards, having incautiously been touched. Mr. Barnes could have found sufficient information about this species in Hume's " Nests and Eggs." Of the beautiful nest and eggs of *Piprisoma agile*, the thick-billed flower-pecker, Mr. Barnes gives no particulars. As it is not in Butler's Guzerat list, I may state that I found three nests at Baroda in last May and June.

The black-headed cuckoo shrike (*V. sykesii*) comes about June 1st, breeds about Baroda in the end of June and beginning of July, and leaves about November. I found four nests last season. The large grey cuckoo shrike (*Graucalus macei*) is a permanent resident here. I found six nests last August near Baroda, each with one egg ; and my men found a nest building in the Police Lines at Khaira, *on the 10th October ;* unfortunately it was destroyed by monkeys.

Mr. Barnes gives no details of the nesting of these two species, though Hume describes both, and Jerdon the latter. Can Mr. Barnes give us any information about the nesting of the white-bellied drongo, *B. cœrulescens* ? It occurs sparingly here between November and April, but seems to go east to the hills to breed.

The Paradise fly-catcher (*M. paradisi*) is very common here during the rains, when it breeds. In all instances except one out of nine nests that I found with eggs last June and July, the birds were in the chestnut plumage, and in that one case the male was white and the female chestnut. The mynas destroyed three nests of one pair of paradise fly-catchers that built in a mango tree near my house. I saw the little fly-catcher defend her first nest for nearly twenty minutes against a myna, that at last retired. Next day, however, the nest was torn to bits, by the myna I suppose. It was twice rebuilt on other branches of the same tree, with the same result. I don't know where she bréd after leaving that tree in disgust.

Mr. Barnes has overlooked the description of the eggs of *Cyornis Tickelli* in *Nests and Eggs ;* and surely to say only of *Dumetia albogularis,* the white-throated wren-babbler, that " it is probably a permanent resident," is to leave out of sight much common information. It is a permanent resident here, and last August I found many nests, which, with the eggs, resembled those described in *Nests and Eggs.*

I may record that 452, *Ixos luteolus,* the white-browed bush bulbul, is common in the ravines along the Mahi and not scarce about Baroda. It seems to prefer the neighbourhood of water and is a hard bird to see, though there is no mistaking its musical trill from some deep thicket. It is not given in Butler's Guzerat list. The Indian oriole, which, Mr. Barnes says, he has found chiefly breeding on *neem* trees, here prefers mango or mhowra trees. I can assure Mr. Barnes that he is quite mistaken in thinking that the Magpie Robin, *C. saularis,* does not remain to breed in Guzerat. It is a permanent resident hereabouts, and I found between May 30th and June 26th last eight nests within a mile of my house. The number of eggs or young varied from two (young) to six (hard set eggs). I have seen the Dayal (a name also given here

to the tailor-bird) in all months here, and have often noticed the
peculiar flurting of the tail over the head, mentioned by Layard (in
Jerdon, who says he has not observed it).

What is a "seasonal visitant," cold season or wet season or hot
season,—who can tell? At any rate *Phylloscopus tristis* is a
"seasonal visitant," while the other *Phylloscopi* are "cold weather
visitants." And why repeat Linnaeus's old misprint of *Anthus
Spinoletta* when naturalists like Prof. Newton give the true form
spipoletta (Yarrell, 4th ed.). And while Mr. Barnes was "at his
larks," he might have told us what was the character of the hind
claw of the genus *Corydalla* (p. 244). Mr. Barnes says the white-eyed
tit (*Z. palpebrosa*) is a common permanent resident in the Deccan,
but that " in other parts of the Presidency it only occurs, I believe,
as a cold weather visitant." I can certify that it breeds here, and
is fairly common, and that I have seen it in nearly every month.
I can also assure Mr. Barnes that *Dendrocitta rufa*, the Indian Tree-
pie, breeds here, and is fairly common all the year round. He says
they become very scarce during the hot weather, and certainly I
have seen great numbers of them then in the hill jungles of Abu and
the Vindhyas, but they do not migrate from our Guzerat plains.
They are very shy and wary birds when breeding, and the nests in
the thick mango foliage are hard to find. As Captain Marshall says
"the eggs of the rose-coloured paster (*P. roseus*) are not known,"
and as Mr. Barnes does not mention them, I may note that a full and
very interesting account of the breeding of these birds is given in the
last edition of Yarrell. I may also say that I kept 18 of them in a
large aviary last season till September, in the hope of their breeding in captivity, but without success.

Estrelda formosa, the green wax-bill, is not very rare hereabouts,
occurring generally in flocks. The common pea-hen I have found
breeding here in the fork of a mango trunk, 10 feet from the ground,
but here, as elsewhere, the usual site is on the ground. Mr. Barnes
ought to have noted that the male of *Turnix taigoor*, the black-
breasted bustard quail, sits on the eggs and minds the babies, while
the female goes round to fight the ladies of the neighbouring families. This brings me up to the grallatores, and I will give only a few
more selections from very many *marginalia* on Mr. Barnes's Handbook.

The lesser florican visits Baroda in small numbers during the
rains, and breeds here; but a few remain here all the year round,
as I have shot them in the following months : February, April, May,

June, July, October. I have seen, but not shot them in August and September. They are not so much reduced by shooting (as Mr· Barnes says) as by snaring. Many are brought in alive to the Camp Bazaar, and sent to me and others as presents, their legs being most cruelly tied with feathers plucked from their own wings. I have released several that had been so tied, and have found that it took several days for them to recover sufficiently for them to leave my garden. The pelican ibis breeds here at Chittral and at Thasra in October. The shell ibis breeds in large numbers, with the white ibis and snake bird, near Khaira. Mr. Barnes says he cannot find any record of the occurrence of the cotton-teal in Guzerat. It is very common, especially in May and June, when there are hundreds on Muwal tank, 20 miles north of Baroda. When the rains fall, they disperse over the country and take up their quarters in some small pond or pool, occasionally

Affording scarce such breadth of brim,
As served the wild duck's brood to swim,

and they nest in the neighbourhood. I extracted a full-sized soft egg from a bird shot near this last September. Mr. Barnes could have found it recorded in Butler's *Gazetteer* list. But enough has been said, I hope, to justify, even from my own very limited experience, the opinion with which I set out, that Mr. Barnes might have got much additional information if he had asked the "Bombay Natural History Society" for it, and might thereby have rendered his book still more deserving than it is at present of being regarded as the standard authority on the birds of the Bombay Presidency.

ON A HYBRID, *OVIS HODGSONI, CUM VIGNEI*, DISCOVERED AND SHOT by Mons. H. DAUVERGNE,

By R. A. Sterndale, f.z.s., &c.

HYBRIDIZATION between the various known species of *Capræ* and *Oves* has been abundantly proved by the instances that have occurred in the London Zoological Gardens. In 1864 and 1865-67 and 1868, a female *Capra Ægagrus*, the Persian Ibex, bore seven kids, the father of which was a Markhor *C. Megaceros*. In 1872 a hybrid between a male *Ovis Aries* and a female *Ovis Musimon*; in 1871 two hybrids between *Ovis Musimon*, the Corsican Moufflon, and our Indian *Ovis Cycloceros* were born ; also in 1871 and 1882 two between the former and *Ovis Aries*. There are two species of deer from Philippine Islands, *Cervus Nigricans* and *Cervus Alfredi* which have twice bred in the gardens, and so have the European and Mesopotamian fallow deer. Sir Victor Brooke in one of his letters to me says he has known the common red deer and the Japanese deer to interbreed. So far the question of interbreeding is amply proved, but the interesting feature of the case is how far is this carried out in the wild state so as to create new species. I am of opinion that, if the truth were fully known, we should have to narrow down our list of goats and sheep. It is an undecided question whether *Ovis Polii* and *Ovis Karelini*, the two great sheep of the Pamir steppes, are not one and the same, and I think that *Ovis Brookei* is the hybrid which forms the subject of this paper. Sir Victor Brooke in a letter to me says : " If we can prove that the form is a hybrid between those two species (*i.e.*, *O. Hodgsoni et Vignei*), it will be much more interesting than if it should prove what is called a distinct species. I do not think the presence of one or even several male *O. Hodgsoni* amongst herds of *O. Vignei* would originate a breed of sheep intermediate in size and character between the two species, the much larger quantity of *Ovis Vignei* blood in the district would, in my opinion, prevail over the infusion of *O. Hodgsoni* blood introduced in such small quantities, and the thus originated larger animals would throw back to the parent stock. If it is a case of hybridization what we should find would be herds of *O. Vignei* with here and there large animals mixing and running with them of *O. Brookei* forms." Now this is exactly what Mons. Dauvergne found. In the mountain range south of the Indus near Zanskar, the precise locality being for obvious reasons withheld from publication, a herd of *Ovis Vignei* were observed for some years to contain a large ram of *Ovis Hodgsoni*, which drove out the weaker Shapoo rams and appropriated the ewes of the herd. He was ultimately one winter killed and eaten by *Chankos* (the Tibetan wolf), but during his stay he produced a family of hybrids possessing greater size of horn and head with characteristic colouring, combining traits of both animals. In course of time these hybrids were crossed again with the *Vignei* stock, and the third generation shows signs of degeneration from the larger sheep and of reversion to the Vignei type.

The skull of the half-bred animals, which the Tartars called *Nyan Shapoo* (the former being the name of the *Hodgsoni* or Ammon

and the latter of the *Vignei*), is nearer in size to *Hodgsoni*, which is double that of the other. The horns of these are rounded in front resembling what has been figured of *Brookei*, but hollowed out behind like *Vignei*. The horns of the quarter-bred are square in front and hollowed behind like the true Shapoo type, but are more massive than the pure-bred Shapoo.*

Now as regards the colour of the skin. The *Nyan* or *Hodgsoni* has no black beard or throat-stripe which *Vignei* has. The half-bred shows no black, but the quarter-bred does in a modified but decided degree. The half-bred turns also in summer to the colour of *Hodgsoni*, having more of a blue grey or lavender tint and less of the fawn colour of *Vignei* with the white throat of *Hodgsoni*, it also gets the dark patch at the side of the neck. The skin of a quarter-bred specimen before me is of a bright fawn above; sides and rump white, and a black stripe down the middle of the throat.

The skull characteristics are as follows :—

	Ovis Hodgsoni	*Half-hybrid*	*Quarter hybrid.*	*Ovis Vignei.*
	Inches.	Inches.	Inches	Inches.
Girth of horn	16½	13½	11¼	10
Length of horns	36	32	22¾	30¼
Length of skull from between horns t tip of premaxillæ	13½	12	9½	9¼
Breadth between orbits	6⅛	5½	4⅝	3⅞
Ditto between frontal sinuses	2¾	2⅞	2⅜	2
Length of teeth	3½	3¼	3	2⅞
Broadest part of palate	2¼	2¼	2	1¾
	80⅝	71⅜	55½	59⅞ 52⅜

In this table there are two noticeable points. It is plain that there is a gradual reversion to the size of *Ovis Vignei*, but although the quarter-bred hybrid has a greater girth of horn than the *Vignei*, the latter has greater length; and this gives it an advantage in all round measurement. Take off these extra 7½ inches in length of horn, and the Shapoo stands at 52⅜ against the quarter-bred's 55½; over 3 inches less. Now comes the question of locality. The nearest Hodgsoni ground to where the Shapoo were located was over sixty miles off, but this is not a barrier to an animal like the Ammon who would cover such a distance in a couple of days. R. A. S.

* I have figured the half-bred horns with rounded fronts on account of their resemblance to the type of *Ovis Brookei*, but I have received another pair of hybrid (half-bred) horns which are quite square in front and as massive as the rounded ones.—R.A.S.

BIRDS' NESTING IN RAJPOOTANA,
(By Lieut. H. Edwin Barnes, D. A. C.)

THESE notes refer only to Neemuch, which, although in Rajpootana, is under the Central Indian Administration.

I was stationed there from December 1883 to the commencement of September 1885, and during the whole time I collected vigorously, but still there are many birds that do undoubtedly breed there that I have overlooked; of these I append a list.

The periods quoted, over which the different breeding seasons extend, were ascertained from personal observation, and represent the time between the earlier and later nests.

2.—*Otogyps calvus* : Scop.

The King Vulture breeds from the middle of February to about the middle of March; some few may breed earlier, but they are exceptions to the general rule, and eggs taken later are generally much incubated.

I took eggs on the 13th and 27th February, and again on the 1st March.

The nests are solitary, and are. huge structures, composed of stout twigs, lined with smaller twigs and leaves, and are generally built in forks of Peepul or other large trees. The egg, there is only one, is oval in shape, measuring 3·52 inches in length by about 2·6 inches in breadth ; the texture is fine, and the shell is very strong. The egg lining is green, but the egg itself is glossless white.

5.—*Pseudogyps bengalensis* : Lath.

The Indian White-backed Vulture breeds much earlier than the King Vulture. I found my first nests on the 9th November, but as three eggs out of five taken on that date contained fully-formed chicks, eggs must have been obtainable much earlier.

They build in colonies, sometimes as many as twenty nests being found on the same tree, and these are at various heights, some being not more than 10 feet from the ground, while others are placed at almost the top of the tree.

Mr. Hume believed that January was the month in which most eggs were laid, but in Neemuch all those I found in December were much incubated, and many eggs had hatched out, so that November in this part of the country would appear to be the best month for nesting. I have never found more than a single egg in any one nest, and this averages somewhat smaller than that of the King Vulture, *viz.*, 3·25 inches in length by about 2·4 in breadth. It is rather coarser in texture. Some eggs are white, but many of them are spotted and blotched with pale reddish brown. The egg lining is a deep green. They are generally much discolored by the droppings of the sitting bird.

6.—Neophron ginginianus : DAUD.

The White Scavenger Vulture breeds about the end of March or commencement of April. The nests are solitary, and are placed in very different situations, on cornices of buildings, edges of rocky or clayey cliffs, and commonly on trees; when in the latter situation, they are not usually built in forks, but are placed on large horizontal branches, or at the junction of a limb with the trunk.

The nest is a large, loose, ragged affair, lined with old rags. The eggs, two in number, are broadish, oval in shape, of a greyish white colour, beautifully streaked, blotched, and clouded with reddish brown. Some are so richly marked as to leave little of the ground colour visible, while others are comparatively plain. They measure 2·62 inches in length by 1·96 in breadth.

11.—Falco jugger : I. E. GR.

The Juggur Falcons breed from the latter end of January to the end of February. They nest indifferently on trees, edges of cliffs, and old buildings; they often appropriate the old nest of a tawny or other eagle. The nest is rather large, cup-shaped if built on a tree, loose and straggling if on a cliff. The eggs, usually four in number, occasionally five, sometimes only three, are nearly perfect ovals in shape, chalky in texture, of a dingy yellowish brown colour, clouded, mottled, and blotched with reddish brown. They measure 2 inches in length by about 1·58 in breadth.

16.—Falco chiquera : DAUD.

The Turumti or Red-headed Merlin breeds during March and the early part of April. The nest is neat, compact, and cup-shaped, and is composed of twigs lined with grass roots. All the nests I have found have been in shady trees, such as Peepul or Banian, and have been fairly well concealed. The eggs, four in number, are exact miniatures of those of the Juggur Falcon. They measure 1·66 inches in length by about 1.26 in breadth.

23.—Astur badius : GM.

The Shikra breeds during April. It takes a very long time to make its nest. I watched a pair for upwards of a month. To-day they would place a few sticks on the nest and to-morrow they would remove them, arranging and re-arranging and taking an infinite deal of trouble, and the result was a nest that would disgrace even a crow. The nests are always built in forks of trees. The eggs, four in number (sometimes only three), are oval in shape, and are of a pure very pale, bluish white colour. They measure 1·54 inches in length by about 1·23 in breadth.

29.—*Aquila vindhiana* : FRANK.

The Indian Tawny Eagle commences to breed about the end of November and nests may be found quite up to the commencement of the hot season, but December and January are the months in which most eggs are laid. The nest is a large structure, composed of stout twigs, lined with green leaves, and it is invariably built upon a high tree. The eggs, two in number, are broadish oval in shape, but are subject to much variation. They are white in colour, more or less spotted and blotched with brown, reddish-brown, and occasionally purple ; they are generally discoloured. The egg lining is sea-green. They measure 2·65 inches in length by about 2·11 in breadth.

38.—*Arcaëtus gallicus* : GM.

I have never succeeded in obtaining an egg of the short-toed eagle, but early in March a native, who often accompanies me in my nesting rambles, reported that he had found a nest on a high tree, with one egg in it. As soon after as convenient, I accompanied him to the spot. There was the nest sure enough, but the egg was gone; the parent birds were hovering round the nest, but they never laid again. The native described the egg as being quite white.

42.—*Haliaëtus leucoryphus* : PALL.

The Ring-tailed Fishing Eagle is another bird whose eggs I failed to procure at Neemuch. I found a nest just finished, at the Panghur Lake, in December, and doubtless I should have obtained eggs had I gone a fortnight later, but the distance was so far, and the road so vile, that I did not think it worth while, as I had a series of eggs which I procured in Sind.

48.—*Butastur teesa* : FRANKL.

The Teesa or White-eyed Buzzard breeds during April. The nest, a rather loose, cup-shaped structure, composed of twigs, unlined, is generally placed in a fork in a mango or other thick foliaged tree. The eggs, three (occasionally four) in number, are broadish ovals in shape, and are delicate pale bluish—or greyish-white in colour, quite devoid of markings. They measure 1·83 inches in length by about 1·54 in breadth.

56.—*Milvus govinda* : SYKES.

The Pariah Kite breeds from early in September quite up to the end of March. I cannot understand how this fact has escaped record, but even Mr. Hume seems to think that Christmas day was an early date to obtain eggs. I have found nests at Abu, Deesa, Hyderabad, Mhow, Poona, Neemuch, and even at Saugor, where I am now stationed, in September. The nest is usually built in a fork, but is sometimes placed on a flat bough. The eggs, two in number (occasionally three), are oval in shape, greyish-white in colour, more or less spotted, streaked, blotched, speckled or clouded with brown and purplish or reddish-brown. Some of the eggs are bright and handsomely coloured, with the markings clearly defined, but others are smudgy and dingily coloured. The nests are more abundant in October and January than at other times, and from this I am led to believe that they have two broods in a year. The egg lining varies from light to deep green, and the eggs average 2·2 inches in length by about 1·78 in breadth.

69.—*Bubo bengalensis* : FRANKL.

The Rock horned Owl breeds during March and April. It makes no nest, the eggs being placed on ledges and in recesses of cliffs, overlooking water. The eggs, three or four in number, are broad oval in shape, and white in colour, with just a perceptible creamy tinge. They average 2·1 inches in length by about 1·73 in breadth.

70.—*Bubo coromandus* : LATH.

The Dusky horned Owl breeds during December and January. They build a large stick nest on trees, which they use for successive seasons, but occasionally they make use of an old Vulture or Eagle's nest. It is usually lined with green leaves. The eggs, usually two in number, vary much both in shape and size, but they are generally broadish oval in shape, and average 2·33 inches in length by about 1·9 in breadth. They are creamy white in colour, and somewhat glossy but coarse in texture.

76.—*Carine brama* : TEM.

The Spotted Owlet breeds from the middle of February to the commencement of April. It nests in holes, and it appears to be a matter of indifference to it whether it be a hole in a tree, a building, a well, an old hay-stack, or even in a rocky cliff. A few leaves and feathers suffice for a nest. The eggs, usually four

in number, are oval in shape, and when fresh and unblown are
of a delicate pink tinge, but are glossless white when much incubated. They measure 1·25 inches in length by about 1 in breadth.
I have often found two pairs of birds using the same hole, and fresh
and incubated eggs are often found together.

84.—*Hirundo filifera* : STEPH.

The Wire-tailed Swallow, to my thinking the handsomest of the
Hirundines, breeds from the latter part of February to April, and
again in August and September. The nest, composed of pellets
of mud, is lined just with a few grass roots, and then with a
plentiful supply of soft feathers. The nest is deep saucer-shaped,
and is placed under the cornice of a bridge, in a niche in a well,
under a culvert, or even under a projecting cliff, always near water.
The eggs, three in number, are longish ovals pointed at one
end, of a glossy white colour, richly speckled with different shades
of reddish brown. They average 0·72 inches in length by about
0·53 in breadth. If the eggs are taken when fresh, the birds will
lay a second, and if these are taken, a third batch in the same nest.

85.—*Hirundo erythropygia* : SYKES.

· The Red-rumped Swallow breeds during the months of June and
July. The nest, composed of pellets of mud, lined with feathers, is
retort-shaped, and is usually built under bridges or culverts, but I
found one nest under a stone slab, projecting over a well. The
eggs, three in number, are pure white ovals, measuring 0·79 inches
in length by about 0·56 in breadth.

89 —*Cotyle sinensis* : I. E. GR.

The Indian Sand Martin breeds during February and March
in holes in banks. These holes, from two to three feet deep according to the nature of the soil, are excavated by the birds themselves. The nest, composed of grass, is well lined with soft
feathers, and contains generally three pure white oval eggs, measuring 0·68 inches in length by 0·48 in breadth.

90.—*Ptyonoprogne concolor* : SYKES.

The Dusky Crag Martin breeds during March and April, and
again in July and August. The nest, composed of pellets of mud,
well lined with feathers, is deep saucer-shaped, and is generally
affixed to the side of a house, under shelter of the eaves. The eggs,
three in number, are white, spotted and blotched with red and
yellowish brown. They measure 0·72 inches in length by about 0·52
in breadth.

100.—*Cypsellus affinis* : I. E. GR.

The Common Indian Swift breeds, I believe, all the year round. The nests are placed under the roofs of verandahs, stables, and such like places, and are composed principally of feathers agglutinated together with saliva. The shape depends altogether on the place in which it is : if in a hole, the nest fits all round it, and necessarily takes its shape ; sometimes it is placed between two rafters, and when these are close together, the nest is long and narrow. Sometimes the nests are isolated, but generally they are built in clusters or congeries. They almost always breed in company. The eggs, three in number, vary much in shape, but are normally very long narrow ovals. They are dead white without any spots. They average 0·87 inches in length by about 0·56 in breadth.

The roof of the verandah of the house in which I lived at Neemuch was literally covered with their nests, so that I had ample opportunities for observing them ; and I believe that there were eggs and nestlings in some or other of them the whole year through.

114.—*Caprimulgus monticolus* : FRANKL.

I found two eggs of Franklin's Night Jar on the 15th June. They were deposited on the bare ground under the scant shelter afforded by a small tuft of grass. They are longish oval in shape, and are of a pinkish cream colour, spotted and blotched with pale brown and faint purple. They measure 1·21 inches in length by 0·84 in breadth.

117.—*Merops viridis* : LIN.

The Common Indian Bee-eater breeds during April. They excavate holes in the banks of nullahs, from two to four feet in extent, according to the nature of the soil. The eggs, four in number, are deposited in the bare soil ; they are nearly spherical in shape and are glossy milk-white in colour. They measure 0·78 inches in length by 0·69 in breadth.

I have often found eggs in the same hole in different stages of incubation.

123.—*Coracias indica* : LIN.

The Indian Roller or Blue Jay breeds during April and May in holes in trees, old walls, or under the eaves of houses. A little grass and a few feathers suffice for a nest. The eggs, four in number, are nearly spherical in shape, and measure 1·3 inches in length by about 1·1 in breadth. They are china-white in colour, and are highly glossy.

129.—*Halcyon smyrnensis* : LIN.

The White-breasted Kingfisher breeds from early in March to the end of May, or even later. It excavates a hole in a river bank, or even in the side of a well. There is no nest. The eggs, five in number (occasionally six), are placed on the bare soil. They are almost spherical in shape, averaging 1·12 inches in length by about 1 in breadth. They are glossy china-white when first laid, but as incubation proceeds, this fades and they become glossless white, and are often discolored.

134.—*Alcedo bengalensis* : GM.

I found but one nesting hole of the little Indian Kingfisher; this was in March, and it contained five unfledged young ones and an addled egg. The egg was nearly spherical in shape, and when fresh must have been of a glossy china-white. It measured 0·79 inches in length by 0·68 in breadth.

136.—*Ceryle rudis* : LIN.

The Pied Kingfishers breed from February to April, unlike the White-breasted Kingfisher. They never make their holes in the sides of wells, but always in river banks over running water. These holes are of great extent, one that I examined extending to quite five feet. The eggs, from four to six in number, are broad ovals, occasionally almost spherical. They are pure china-white when blown, and are highly glossy. They measure 1·2 inches in length by about 0·91 in breadth.

148.—*Palæornis torquatus* : BODD.

The Rose-ringed Paroquet breeds from the end of February to early in April. It nests in holes, generally in trees, but occasionally in buildings and old walls. The eggs, usually four in number, are broadish ovals in shape, pointed at one end, and are of a pure glossless white. They measure 1·22 inches in length by about 0·95 in breadth.

197.—*Xantholæma hæmacephala* : P. L. S. MÜLL.

The Coppersmith begins to breed in February, and eggs may be found quite up to the middle of April, but most of them are laid in the commencement of March. They select a branch decayed internally, and into this they cut a small circular hole; there is no nest. The eggs, three or four in number, are long, narrow, pure white ovals, measuring 1 inch in length by about 0·7 in breadth.

212.—*Coccystes jacobinus* : BODD.

I never obtained an egg of the Pied-crested Cuckoo at Neemuch that I could be quite sure of, but then the bird is comparatively rare, but at Mhow, where the bird literally swarms during the monsoon, I obtained an egg extracted from the oviduct of a female.

214.—*Eudynamis honorata* : LIN.

The Koel lays her eggs in nests of the Common Crow, usually one egg in a nest, occasionally two, but I once found three, but as these eggs differed from each other, I am inclined to think they must have been the produce of different birds. I have never found the Crow eggs broken. The eggs vary much both in colour and size, pale sea-green, oily-green, dull olive-green and dingy stone-coloured varieties all occur, and the markings are olive or reddish brown and dull purple. They average 1·2 inches in length by 0·92 in breadth.

217.—*Centrococcyx rufipennis* : ILL.

The Crow Pheasant or Coucal breeds from May to July, or even later. It builds a large, irregular, domed, globe-shaped nest, composed of twigs and coarse grass, lined with leaves. The nest is placed in the centre of a thorny thicket or tree. The eggs (I have never found more than three) are broad, white, chalky ovals, measuring 1·43 inches in length by about rather less than 1·17 in breadth.

234.—*Cinnyris asiatica* : LATH.

The Common Purple Honeysucker commences to breed in March, and nests may be found quite up to the beginning of the rains. The nest is pendant-shaped, something like a Florence flask, or oval with a tapering neck. This is suspended from the end of a slender branch or twig. All sorts of material are made use of in constructing the nest: fibres, cobwebs, hair, fine grass, bits of straw, lichens, dead leaves, flower petals, pieces of rag, &c., are all pressed into service and are neatly and compactly woven together. It is well lined with soft vegetable down. The nest at a short distance resembles one of the bunches of cobwebs, so commonly met with on trees and bushes. The entrance, which is on one side, about half way up, is shaded by a canopy, beautifully adapted to keep out the rains. The eggs, two or three in number, are dingy little ovals. The ground colour is greenish or greyish-white, usually almost obscured by greyish-brown or purplish-grey ill-defined markings. They average 0·64 inch in length by about 0·46 in breadth.

256.—*Lanius lahtora*: SYKES.

The Indian Grey Shrike breeds from March to early in July, but the favorite month seems to be April, as I have found many more nests in that month than in any other. The nest is generally placed

in the centre of a thorny bush or small tree, and is composed of
various materials, such as thorny twigs, coarse grass, pieces of rag,
&c., which form the body of the nest, while the interior is lined with
fine grass, hair, and the like.   The eggs, usually four in number, are
broad oval in shape, pointed at one end, and are greenish-white
in colour, with brown and purple markings ; sometimes these are ill
defined, but occasionally they stand out clear and distinct, and not
seldom form an irregular zone at the larger end.   They measure
1·05 inches in length by about 0·8 in breadth.

257.—*Lanius erythronotus* : Vig.

The Rufous-backed Shrike breeds from June to August.   The nest
is similar to that of *L. Lahtora*, but is perhaps, as a rule, more compactly built.   The eggs, too, are similar in all respects except size,
measuring 0·92 inch in length by rather more than 0·7 in breadth.

260.—*Lanius vittatus*: Valenc.

The Bay-backed Shrike breeds from March to July.   The nest,
placed in a fork of a small babool tree, is deep cup-shaped, neatly
and compactly built, and is composed of fine twigs, grass roots, &c.,
lined with feathers and fine grass.   The eggs, four in number, are
broad ovals in shape, and are of a pale greyish or greenish-white
colour, with an ill-defined zone of brownish and purplish spots at the
larger end with a few spots of the same colour scattered over the
remaining surface.   They measure 0·83 inch in length by about
0·65 in breadth.

276.—*Pericrocotus peregrinus*: Lin.

The Small Minivet breeds during July and August.   The nest
is small, neatly and compactly built, of a deepish cup-shape, and
is generally located in a fork of a branch of a tree at some height
from the ground.   It is composed of fine twigs bound together with
cobwebs, and so closely resembles the bark of the tree, that it looks
like a mere knot or excrescence ; there is very little lining.   The
eggs, three in number, are rather broadish ovals, of a pale greenish-
white colour, speckled, spotted and blotched with bright brownish-red.
They measure 0·66 inch in length by about 0·53 in breadth.

278.—*Buchanga atra*: Herm.

The King Crow breeds during May and June.   A few nests may
be found in July, but by far the greater number are to be found
during the latter part of May and the commencement of June.   The
nests are built in forks at the extremities of branches, generally at

some considerable height from the ground. They are strongly but slightly made, so much so, that the contents of the nest can be seen from below; they are composed of grass stems and roots neatly interlaced. The eggs, four in number, are glossless white with numerous spots and specks of rusty red and reddish-brown; occasionally the eggs are of a deepish salmon tint, the spots and specks being brownish-red. I have never found a pure white egg. They measure one inch in length by about three quarters of an inch in breadth.

288.—*Muscipela paradisi* : Lin.

The only nest of the Paradise Flycatcher that I found was in June, and it was not quite finished. I sent a shikaree a week later to examine it, when it contained a single egg which he brought in; it measures 0·8 inch in length by 0·6 in breadth, and is an exact miniature of a richly coloured King Crow's egg.

292.—*Leucocerca aureola*: Vieili,

The White-browed Fantail breeds from the latter part of February to the commencement of August, but most nests are found in March and July, and from this I infer that they have two broods in the year. The nest is usually placed on the upper surface of a horizontal branch; it is round and cup-shaped, rather deep, and is composed of fine grass roots, tightly bound with cobwebs, and is a very beautiful nest, not much bigger than the top of a wine-glass. The eggs, three in number, are little buffy ovals, with a nimbus or belt of spots round the middle. They measure 0·66 inch in length by about 0·5 in breadth.

385.—*Pyctoris sinensis* : Gm.

The Yellow-eyed Babbler breeds from July to September. The nests are placed either in small forks in trees, or between the stalks of growing corn or sedges. When in the former situation, the nest is deep cup-shaped, but in the latter it is more cone-like, the bottom of the nest being frequently prolonged to a point. The nest is very handsome, and is composed of broad-leaved grasses, strips of bark, vegetable fibres and cobwebs. The eggs, four in number, vary much in colour, some being white with bold hieroglyphic blotches of rusty red and reddish-brown ; others are pinkish-white, but so closely stippled and streaked with bright brick-dust red as to leave little of the ground colour visible. Every possible combination of these two types is to be met with, but all the eggs in a nest are of the one kind. In shape they are broadish ovals, but here again considerable variations occur.   They measure 0·73 inch in length by about 0·6 in breadth.

432.—*Malacocercus terricolor* : HODGS.

I only came across one undoubted nest of the Bengal Babbler. This was in April, and it contained four eggs.   Both nest and eggs are absolutely indistinguishable from those of a *Malcolmi*.

436.—*Argya malcolmi* : SYKES.

The Large Grey Babbler is very common, and I have found nests in each month from January to December. They have, I believe, several broods in the year, and even when nesting associate in small parties of seven or eight.   The nests, composed of grass roots, are loosely but neatly woven together, and are placed amongst the smaller branches of babool trees, at no great height from the ground.   The eggs, four in number, are rather broadish ovals, of a very glossy greenish-blue colour.   They measure 1 inch in length by about 0·78 in breadth.

438.—*Chatarrhœa caudata* : DUM.

The Striated Bush Babbler breeds from March to July.   The nest is usually placed in a low thorny bush, and is composed of grass roots and stems ; it is deep cup-shaped, neatly and compactly built. The eggs, three or four in number, are longish ovals, slightly compressed at one end, and are of a pure, pale, spotless, blue colour. They measure 0·85 inch in length by about 0·64 in breadth.

462.—*Molpastes hæmorrhous* : GM.

The Common Madras Bulbul breeds from April to September. Nests are occasionally found even earlier than this, but they are exceptions to the general rule.   The nest is usually placed in a fork in a bush or small tree, and is of a neat cup-shape, composed of grass, roots, &c., lined with hair, fine grass and fibres.   The eggs, three or four in number, are normally longish ovals, slightly pointed at one end, and vary very much in colour. One type is pinkish white, thickly speckled and stippled more or less over the whole surface with blood red ; in another type, the ground colour is pink with large blotches of deep red and smaller ones of inky-purple.   Between these two types almost every combination occurs.   They measure 0·9 inch in length by about 0·65 in breadth.

468.—*Iora tiphia* : LIN.

The White-winged Green Bulbul or Iora breeds at Neemuch in April and August.   I only found two nests : one in April contained three unfledged nestlings, and the other in August contained three fresh eggs. In both instances the nests were placed in forks of guava trees, and were neatly and strongly yet slightly built, composed of grass roots and fibres bound together with cobwebs. The eggs are broadish ovals in shape, and are creamy white in colour, with long streaks of purplish and yellowish-brown. They measure 0·69 inch in length by 0·55 in breadth.

470.—*Oriolus kundoo* : SYKES.

The Indian Oriole breeds during July and August. The nest, pocket-shaped, is suspended between a fork at the extremity of a branch of a large tree. It is composed of grass and roots, bound round the twigs forming the fork with strips of bark ; it is lined with fine grass. The eggs, three in number, are longish ovals, pointed at one end, and are of a beautiful glossy china white colour, with clearly defined, deep, blackish-brown spots. They measure 1·1 in inches length by about 0·8 in breadth.

475.—*Copsychus saularis* : LIN.

The Magpie Robin breeds during April and May. The nests are placed in holes in trees or old stone walls, and are often mere pads, with a depression in the centre for the reception of the eggs, and are composed of roots, grass, hair, &c. The eggs, four in number, are moderately broad ovals, pointed at one end, and are bluish or greenish-white in colour, speckled and spotted with different shades of reddish-brown. They measure 0·81 inch in length by about 0·67 in breadth.

480.—*Thamnobia cambaiensis* : LATH.

The Northern Indian Robin breeds from March to the middle of July. The nest is placed in a hole in a tree or stone wall, under a bank or the eaves of houses, and such like places, and is generally a mere pad, composed of roots, grass, hair, leaves, feathers, &c. The eggs, four in number, are oval in shape, pointed at one end, and are pale greenish-white in colour, speckled and spotted with different shades of reddish-brown.

494.—*Cercomela fusca* : BLYTH.

The Brown Rock Chat breeds from March to the end of July, rearing, I believe, two or three broods in the season. The nests, which are mere pads of grass roots and hair, are placed in holes in stone walls, in clefts in rocks, and under banks. The eggs, three or four in number, are broadish ovals pointed at one end, and are of a pure, pale, blue colour, with spots and specks of red and reddish-brown, chiefly confined to the larger end, where they often form a belt. They measure 0·82 inch in length by about 0·62 in breadth.

530.—*Ortholomus sutorius* : FORST.

The Indian Tailor Bird breeds from July to the end of September. The bird selects a largish leaf and manages to fasten the edges together by a few shreds of cotton, and in the cavity thus formed it constructs a nest, composed almost exclusively of cotton, with only just sufficient hairs in it to give it elasticity and to keep it in shape. This is the most common type of nest, but often they sew two or more leaves together. The eggs, three in number, are longish ovals, generally whitish with a few blotches of bright rusty red. Occasionally the eggs are pale greenish-white with the rusty red markings less bright. They measure 0.64 inch in length by about 0.45 in breadth.

534.—*Prinia socialis* : SYKES.

The Ashy Wren Warbler breeds about the same time and in a somewhat similar manner to the Tailor Bird, but the nest is not so neatly made, and grass and fibres are oftener used in its construction. The eggs, four in number, are broadish ovals of a glossy brick red or mahogany colour. They measure 0.64 inch in length by about 0.47 in breadth.

535.—*Prinia stewarti* : BLYTH.

Stewart's Wren Warbler breeds in a precisely similar manner to the Ashy Wren Warbler, and I could discover no constant difference either in the shape, size, or colour of the eggs.

543.—*Drymœca inornata* : SYKES.

The Earth-brown Wren Warbler breeds during the monsoon, that is, from July to the end of September. A favourite site for a nest is under the broad leaf of a shrub that grows very commonly in the district. It constructs a purse-shaped nest, with fine shreds or strips of grass. The leaf which forms a roof to the nest is pierced through and through with these shreds, and here and there a strip of grass is fastened to an adjoining leaf or stalk. Another common type of nest is formed by attaching strips of grass to thorny twigs, so as to form a sort of framework, and then carefully weaving other strips between them, the nest necessarily taking the shape of the framework. Another kind of nest is simply a rather less neatly woven purse, attached to the stems of growing corn or sedges. The nest is never lined. The eggs, four, sometimes five, in number, are oval in shape, and glossy pale greenish-blue in colour, with blotches and spots of deep chocolate and reddish-brown, and an intricate tracery of closely interlaced delicate lines round the large end ; occasionally these lines are absent. They measure 0.6 inch in length by about 0.45 in breadth.

545.—*Drymœca sylvatica* : JERD.

The Jungle Wren Warbler breeds during the monsoon, making a globular nest of grass and fibres. The eggs, four or five in number, are of two distinct types, pale greenish-white with very close but minute specks of rusty red, and white with similar markings. They measure 0·69 inch in length by about 0·5 in breadth.

551.—*Franklinia buchanani* : BLYTH.

The Rufous-fronted Wren Warbler breeds during July, August, and the early part of September. The nest, composed of grass, is loosely constructed, and is placed in low bushes or scrub. The eggs, five in number, are broadish oval in shape, white in colour (tinged bluish), thickly and finely speckled with dingy red. They measure 0·61 inch in length by about 0·48 in breadth.

589.—*Motacilla maderaspatenis* : GM.

The Pied Wagtail breeds during March, April and May. The nest is a mere pad of grass, roots, hair, &c., placed in a hole in a wall or well, on a rocky or earthy ledge, or anything solid, but always in the vicinity of water. The eggs, three or four in number, are broadish oval in shape, pointed at one end, and are greenish or earthy-white in colour, with dingy brown markings. They measure 0·9 inches in length by about 0·65 in breadth.

660.—*Corvus macrorhynchus* : WAGL.

The Bow-billed Corby breeds from the latter end of February to about the middle of April, making the usual corvine stick nest. The eggs, four in number, are moderately broad ovals in shape, and are greenish-blue in colour with spots, streaks, and dashes of sepia, blackish and olive-brown. They measure 1·73 inches in length by about 1·19 in breadth.

663.—*Corvus splendens* : VIEILI.

The Ashy-necked or Common Indian Crow breeds during May and June. The eggs are of the usual corvine type, but are much smaller than those of the Corby, measuring 1·4 inches in length by about 0·98 in breadth.

684.—*Acridotheres tristis* : LIN.

The Common Myna breeds during June and July. A favourite spot for a nest is on the top of a pillar, in a verandah, just under the roof, but holes in trees and walls are not neglected. The nest is a mere collection of fine twigs, roots and grasses. The eggs, four or five in number (quite as often one as the other), are longish ovals in shape, and unspotted greenish-blue in colour. They measure 1·2 inches in length by about 0·86 in breadth.

685.—*Acridotheres ginginianus* : LATH.

The Bank Myna breeds in holes, made by themselves, in river banks, about May. The eggs, four in number, are counterparts of those of the Common Myna, but are smaller. They measure 1·05 inches in length by about 0·87 in breadth.

687.—*Sturnia pagodarum* : GM.

The only nest of the Brahminy Myna that I found was in June ; it was in a hole in a tree, and contained three fresh eggs. They are longish ovals in shape, and are of a pale greenish-blue colour, and measure 0·97 inches in length by about 0·73 in breadth.

694.—*Ploceus philippinus* : LIN.

The Baya or Weaver Bird commences to breed about the latter end of July, that is, when the rains have set in ; it is a gregarious builder, as many as forty nests being frequently counted upon one tree, which is usually a thorny babool, growing over water, river, tank or well, it does not matter which, thus obtaining greater protection. The nests are retort-shaped, and are composed of strips of grass, ingeniously interwoven ; the grass is always used green. They commence operations at the extreme end of a slender twig, and for the first few inches the nest is solid, gradually increasing in size. After about a foot of the nest is made, they commence to form a receptacle for the eggs on one side and a tubular entrance opposite, a strong loop being made across the nest to form the division. The egg compartment is about seven inches in length by six in breadth and four and-a-half in width, but they vary much. The above dimensions are of a very fine nest. The tubular entrance is generally five or six inches in length, but as the male bird goes on increasing the length during the time the female is sitting, it often reaches an almost incredible length. I have seen one measuring sixteen inches. I am puzzled as to what the ordinary number of eggs is. I have often found two eggs, much incubated ; many times I have met with four, and on one occasion I took seven from the same nest. The eggs are moderately long ovals, pointed at one end, and are dull white in colour. They measure 0·82 inches in length by about 0·59 in breadth.

695.—*Ploceus manyar* : HORS.

The Striated Weaver Bird breeds about the same time as its relative *P. Philippinus*. The nest is very similar, but instead of being affixed to the end of a bough, it is fastened to the top of a bunch of reeds growing in water. The eggs are much like those of *P. Philippinus*, but are rather smaller.

703.—*Amadina malabarica* : LIN.

I have found nests of the Pintail Munia throughout the year. They are usually placed in low thorny bushes, but they are very variable in the site they select. I once found a nest under the eaves of an out-house, and not unfrequently they make their nests in the sticks forming the foundation of a Kite's nest. The eggs, pure white in colour, vary from 5˙to 9 in number, but I am inclined to think that occasionally more birds than one lay in the same nest. They measure 0·6 in length by about 0·47 in breadth.

704.—*Estrelda amandava* : LIN.

I found but a single nest of the Red-waxbill, and it containe l four half-fledged nestlings. This was in October.

706.—*Passer domesticus* : LIN.

The House Sparrows breed from February to August, and are quite a nuisance the while ; no amount of persecution seems to deter them from building in a place when once they have made up their minds to it.

711.—*Gymnoris flavicollis* : FRANKL.

The Yellow-throated Sparrow breeds during April and May in holes in trees. The eggs, four in number, are much smaller and darker ·than those of *Passer domesticus*. They measure 0·74 in length by 0·54 in breadth.

756.—*Mirafra erythropygia* : JERD.

The Red-winged Bush Lark breeds from March to September. I am inclined to think that it has two broods in the year, as nests are much more commonly found in March and April, and again in August and September. The nest is built upon the ground, under the shelter of a tussock of grass, and is composed of grass stems and roots. The eggs, four in number, are oval in shape, and are of a greenish-white colour, speckled and spotted with various shades of reddish and yellowish-brown. They measure 0·78 inches in length by about 0·6 in breadth.

757.—*Mirafra cantillans* : JERD.

The Singing Bush Lark is decidedly rare at Neemuch, and I only succeeded in finding one nest, which was in September. This was similar to that of the Red-winged Bush Lark as regards locality and material, but was more perfectly domed over. The eggs, four in number, were much incubated. They measured 0·78 inches in length by about 0·6 in breadth.

758.—*Ammomanes phœnicura* : FRANKL.

The Rufous-tailed Finch Lark breeds during March and April ; the nest is a mere circular pad, placed in a cavity under a clod of earth, and is composed of grass roots, scantily lined with a few hairs ; the eggs, usually three in number (I once found four), are very variable in size, shape and color, but are usually longish ovals, measuring 0·85 inches in length by about 0·62 in breadth, and are usually yellowish-white in color, with specks and spots of reddish or yellowish-brown·

760.—*Pyrrhulauda grisea* : SCOP.

I found nests and eggs of the Black-bellied Finch Lark in each month throughout the year, with the exception of July and August. The nest, which is a soft pad, with a depression for the eggs, is placed in a footprint or slight hollow in the ground, under the shelter of a clod of earth or tussock of grass. The eggs, two in number, occasionally three, are moderately long ovals, of a dingy or greyish-white color, thickly speckled, sprinkled and spotted with yellowish-brown. They measure 0·73 inches in length by about 0·55 in breadth·

765.—*Spizalauda deva* : SYKES.

The Southern Crown-crest Lark breeds during July, August and September ; the nest is placed on the ground in the centre of, or under the shelter of, a tussock of grass, and is composed of grass roots and fibres ; it is of a shallow cup-shape. The eggs, two or three in number, quite as often one as the other, are oval in shape, pointed at one end, and are of a dingy white colour, profusely spotted and speckled with yellowish and earthy brown. They measure 0·86 inches in length by about 0·63 in breadth.

767.—*Alauda gulgula* : FRANKL.

The Indian Sky-lark breeds during the month of July, possibly both earlier and later, but July is the only month in which I have obtained eggs. The nest, composed of fine grass, is placed in a depression in the ground, and the eggs, three or four in number, are moderately broad ovals, of a dingy or greyish-white colour, spotted and speckled with yellowish-brown and purplish-grey. They measure 0·8 inches in length by about 0·6 in breadth.

773.—*Crocopus chlorigaster* : BLY.

I found the Southern Green Pigeon breeding in March. The nest, which was of the usual stick type, contained two pure white eggs. They were much incubated, but were still highly glossy. They measured 1·2 inches in length by 0·9 in breadth.

788.—*Columba intermedia*, BRICKL.

By far the favourite site for the nest of the Indian Blue Rock Pigeon is in holes in masonry wells.

794.—*Turtur senegalensis* : SMIL.

The Little Brown Dove breeds throughout the year; it shows a decided preference for prickly-pear bushes, as I found twenty nests in them to one elsewhere. The eggs average an inch in length to about 0·84 in breadth.

795.—*Turtur suratensis* : GM.

The Spotted Dove has not such an extensive breeding season as the Little Brown Dove; indeed, I have only found nests in September. The eggs measure 1·1 inch in length by about 0·85 in breadth.

796.—*Turtur risorius* : LIN.

The Common Ring Dove breeds from October to July; at least I have taken eggs in each of these months, but I believe that had I searched, I should have found them during the remaining months. The eggs measure 1·15 inches in length by about 0·92 in breadth.

797.—*Turtur tranquebaricus* : HERM.

I only found nests of the Ruddy Ring Dove in November, so that its breeding season seems much more restricted than is generally the case with doves. The bird is not common, and is very locally distributed. The eggs measure 1·01 inches in length by about 0·8 in breadth.

800.—*Pterocles fasciatus* : SCOP.

I was very unfortunate in not obtaining eggs of the Painted Grouse. The birds are by no means uncommon, and I have frequently obtained young ones.

802.—*Pterocles exustus* : TEM.

The Common Sand Grouse has a very extended breeding season, as I have found eggs from January to June. They are three in number; and are placed in a depression in the soil, and are of a long cylindrical shape, equally rounded at both ends. They are of a greenish-stone colour, spotted, streaked, clouded and blotched, olive-brown and pale inky purple. They measure 1·45 inches in length by about an inch in breadth.

803.—*Pavo cristatus* : LIN.

The Pea-fowl breeds during August and September, when the rains are at their height. The eggs, six or seven in number, are laid in a depression in the soil (scratched by the hen), scantily lined with a few grass stems or leaves. They are broadish ovals, slightly pointed at one end, and are creamy-white or pale cafe-au-lait in colour, pitted all over like a Guinea-fowl's egg. They measure 2·75 inches in length by about 2 in breadth.

814.—*Galloperdix spadiceus* : VAL.

The Red Spur Fowl breeds during June and July, and probably earlier, as I saw a brood of chicks early in July that must have been hatched in the beginning of June. The nest is very slight, placed in a depression in the ground, scratched by the hen herself. The eggs, from four to six in number, are miniatures of those of the domestic fowl. They measure 1·6 inches in length by about 1·2 in breadth.

819.—*Francolinus pictus* : JAR. & SEL.

The Painted Partridge lays after the rains have well set in, *viz.*, about August and September. The nest is a very loosely made pad, placed in a depression in the ground. The eggs, six or seven in number, are peg-top shaped, and are of a smoky white colour. They measure 1·4 inches in length and about 1·15 in breadth.

822.—*Ortygornis pondiceriana* : GM.

The Grey Partridge breeds from the end of March to quite the middle of June. The eggs are occasionally found on the bare ground, but there is generally a more or less compact pad, placed in a depression in the ground under cover of a tuft of grass. The eggs, six to nine in number, are slightly elongated ovals, pinched in more or less at one end, and are of a slightly soiled white colour, and measure 1·3 inches in length by about 1 inch in breadth.

826.—*Perdicula asiatica* : LATH.

The Jungle Bush Quail lays towards the end of the rains. I have never succeeded in obtaining eggs, but have many times flushed broods of chicks.

827.—*Perdicula argoondah* : SYKES.

The Rock Bush Quail breeds from August to December. They may commence earlier, but I have only found eggs in the months mentioned. The nest is placed in the ground generally under a clump of grass or shrub, and is composed of a few blades of grass. The eggs, six or seven in number, are much like those of the Grey Partridge, but are much smaller. They measure 1 inch in length by about 0·82 in breadth.

836.—*Eupodotis edwardsi* : GRAY.

The Indian Bustard is fairly common at Neemuch. I have an egg that was found on the bare ground under a tuft of Sarpat grass in July. This egg is of a dark olive brown colour, with a few streaks and smudges of a darker shade. It measures 3·1 inches in length by 2·25 in breadth.

839.—*Sypheotides aurita* : LATH.

The Likh or Lesser Florican does not breed until the rains have well set in, that is, not until September and October. There is no nest ; the eggs, three or four in number, being deposited on the bare ground, under cover of a stunted bush or tussock of grass. They are broad oval in shape, and are of an olive green colour with reddish brown streaks and smudges. They measure 1·9 inches in length by 1·6 in breadth.

840.—*Cursorius coromandelicus* : GM.

The Indian Courser or Courier Plover breeds during March and April. There is no nest. The eggs, two or three in number, are deposited on the bare ground, under shelter afforded by a clod of earth or tussock of grass. Owing to their colour assimilating so closely to the ground on which they are placed, they are very difficult to find. The eggs are nearly spherical in shape, and are of a yellowish stone colour, closely spotted, speckled and lined with blackish brown, and having a few underlying clouds or smudges of pale inky grey. They measure 1·2 inches in length by 0·98 in breadth.

850.—*Ægialitis minutus* : PALL.

The Lesser Ringed Plover breeds abundantly during March and April. There is no nest. The eggs, three in number, are placed on the sand, in the bed of a river ; they are broad oval in shape, much pointed at one end, and are of a yellowish stone colour, thinly lined and spotted with blackish brown. They measure 1·2 inches in length by about 0·83 in breadth.

The anxiety exhibited by these little Plovers, when they have young, and their many devices to entice intruders away from their vicinity, quite equals anything recorded of the Lapwing. On the 17th April, while wandering on the banks of a nullah, my attention was arrested by the peculiar movements of one of these birds. It was lying on its side as if in death agony with its wings fluttering and quivering; it would make an attempt to fly, but after proceeding a yard or two it would fall down headlong as if shot. Suspecting that it had eggs or young near, I made a diligent search, but could find nothing, the bird all the time accompanying me and making the most frantic efforts to distract my attention. I left off searching, but carefully watched the bird from a distance. After a short time it settled itself down, as a hen would squatting over chicks. I carefully marked the spot, made a sudden rush at it, and then on my hands and knees I carefully felt all round, and presently found a tiny fluffy chick, apparently stone dead. I thought that I must have stepped upon it and killed it. I felt very sorry, but all at once I saw the little beggar open one eye and take a look at me. I placed it on the ground, and taking my eye off of it for a moment, it disappeared; and it was only after a long and painstaking search that I again found it, still apparently dead. I moved a few paces away and watched it. After a moment it opened its eyes, gave a slight stretch, and disappeared as if by magic. I found three broken egg shells close by, and they appeared as if the chicks had only just been hatched, and there must have been two others close by me, although they escaped my search.

855.—*Lobivanellus indicus* : BODD.

The Red-wattled Lapwing breeds from April to July. There is no nest. The eggs, four in number, are placed on the ground, almost always in the vicinity of water. They are broad oval in shape, much pointed at one end, or I should say a peg-top shape. They vary somewhat in colour, but are usually of a yellowish buff, blotched and streaked with reddish brown. They measure 1·64 inches in length by about 1·25 in breadth.

856.—*Lobipluvia Malabarica* : BODD.

The Yellow-wattled Lapwing breeds during April and May. There is no nest. The eggs, four in number, are deposited on the bare ground, without any attempt at concealment; they are not partial to water, but frequent by preference bare sandy plains. The eggs are similar in shape to those of *Lobivanellus indicus*, but are much smaller, only measuring 1·45 inches in length by 1·06 in breadth.

862.—*Grus antigone* : Lin.

The Sarus breeds freely during August and September, but I found two fresh eggs in February while duck shooting and two incubated in March, probably both these clutches belonged to birds that had had their first eggs accidentally destroyed. The eggs, two in number, are of an elongated oval shape, pointed at one end. They vary in colour, but are generally creamy white, more or less spotted and blotched with pale yellowish-brown and purplish-pink. They measure 3·9 inches in length by 2·55 in breadth.

873.—*Rhynchœa bengalensis*: Lin.

I f und the Painted Snipe breeding in May. It probably breeds both earlier and later than this, but this was the only month in which I obtained eggs. They are broadish oval in shape, pinched in at one end, and are of a buffy colour, blotched and streaked with rich black brown. They measure 1·4 inches in length by 1 in breadth.

900.—*Metapodius indica* : Lath.

The Bronze-winged Jacana breeds during July and August, making a floating nest of weeds. The eggs (I never found more than four but then they were all fresh), are broad ovals, pointed at one end, and are generally of a rich cafe-au-lait colour, but are subject to considerable variation. One clutch I have is a dark olive brown, while another is a very pale stone brown. The eggs of this last clutch are abnormally small. The markings, consisting of a network of entangled lines, are very deep blackish brown. The eggs are highly glossy, and measure 1·47 inches in length by 1·02 in breadth.

901.—*Hydrophasianus chirurgus*: Scop.

The Pheasant-tailed Jacana breeds during August and September. The nest is a floating one, composed of grass and aquatic plants. The eggs, four in number, are peg-top shaped, and are of a glossy rufous or greenish bronze. They measure 1·46 inches in length by about 1·1 inch in breadth.

902.—*Porphyrio Poliocephalus*: Lath.

The Purple Coot breeds during September. The nests, built of rushes and reeds, are floating but not free, and occasionally they rest upon the ground. The eggs, seven or eight in number, are broadish ovals in shape, and are of a pale pinkish stone colour, thickly spotted and blotched with rich red brown and pale purple. They measure 1·93 inches in length by about 1·4 in breadth.

903.—*Fulica atra*: LIN.

I did not succeed in finding a nest of the Common Coot, but a native fisherman, who has often given me information regarding nests and eggs, and whom I have generally found reliable, reported that he had seen a batch of newly-hatched chicks in April. I was too busy at the time to go out, so could not verify his statement, but suspect that what he saw was a brood of the white-breasted Water Hen.

907.—*Erythra phœnicura*: PENN.

The White-breasted Water Hen breeds from May to August. All the nests I have found have been placed in the branches of dense bushes or trees close to water. The eggs, four in number, differ much in size, shape and colour. Eggs of the same clutch will even differ. The usual type is creamy white, with yellowish brown and light red spots and blotches, with apparently underlying markings of pale bluish gray. Some eggs I have are white with scarcely any markings. They are usually broadish oval in shape, and average 1·55 inches in length by about 1·18 in breadth.

930.—*Ardeola grayi*: SYKES.

The Indian Pond Heron breeds from June to August, generally in small colonies, but isolated nests not unfrequently occur. They are composed of sticks, and are of a platform shape. The eggs, four or five in number, are rather longish ovals, slightly pointed at one end, and are of a deep sea-green colour. They measure 1·48 inches in length by 1·17 in breadth.

938.—*Tantalus leucocephalus*: GMEL.

The Pelican Ibis breeds in colonies during March and April. The nests are small, rough platforms, composed of sticks, and are placed high up in lofty trees, often in the vicinity of villages. The eggs (I never found more than four, but they were fresh and probably the birds lay more) are elongated ovals, pointed at one end, and are of a dull unspotted white. They measure 2·77 inches in length by about 1·88 in breadth.

950.—*Sarcidiornis melanonotus*: PENN.

I have been very unfortunate with the Nukhtah, as I could never obtain an egg, but several times in September I have shot half-fledged young.

951.—*Nettapus coromandelicus* : LIN.

This is another bird whose eggs I have been unable to procure, although I have often seen the young.

952.—*Dendrocygna javanica* : HORS.

The Whistling Teal breeds during August and September. In Neemuch I have never found the nests on trees, but always amongst the sedges on the border of a tank. The eggs, six or seven in number, are broad oval in shape, and are milky white in colour. They measure 1·85 inches in length by about 1·49 in breadth.

959.—*Anas pœcilorhyncha* : CUV.

I have never succeeded in obtaining the eggs of this duck, but have often at the end of the rains shot the ducklings.

975.—*Podiceps minor* : GM.

The Dabchick breeds during September and October. The nest, a floating one, is composed of aquatic weeds and sedges. The eggs, four or five in number, are, when freshly laid, chalky white, but as incubation proceeds they become much stained, from the habit the bird has of covering her eggs with wet weeds when she leaves the nest. They are elongated ovals in shape, pointed at each end, and measure 1·39 inches in length by about 0·99 in breadth.

985.—*Sterna seena* : SYKES.

I found four eggs of the Large River Tern in the sandy bed of the river in May. They are broad ovals in shape, and are of a pale greenish-grey colour, blotched and streaked with brown, and having underlying clouds of a pale inky purple. They measure 1·65 inches in length by about 1·26 in breadth.

In addition to the above, of which I have either procured eggs or seen the young, the following birds must, I am sure, breed at Neemuch, as I have constantly noted them throughout the year :—

55.—Haliastur indus : Bodd.

57.—Pernis ptilorhynchus : Tem.

59.—Elanus melanopterus : Daud.

65.—Syrnium ocellatum : Lesson.

104.—Dendrochelidon coronata : Tick.

107.—Caprimulgus indicus : Lath.

144.—Ocyceros birostris : Scop.

147.—Palæornis eupatria : Lin.

149.—Palæornis purpureus, P. L. S. Müll.

160.—Picus mahrattensis : Lath.
180.—Brachypternus aurantius : Lin.
219.—Taccocua leschenaulti : Less.
265.—Tephrodornis pondicerianus : Gm.
600.—Corydalla rufula : Vieill.
645.—Parus nipalensis : Hodgs.
647.—Machlolophus xanthogenys : Vig.
674.—Dendrocitta rufa : Scop.
696.—Ploceus bengalensis : Lin.
830.—Coturnix coromandelica : Gm.
832.—Turnix taigoor : Sykes.
834.—Turnix joudera : Hodgs.
835.—Turnix dussumieri : Tem.
905.—Gallinula chloropus : Lin.
908.—Porzana akool : Sykes.
917.—Xenorhynchus asiaticus : Lath.
923.—Ardea cinerea : Lin.
924.—Ardea purpurea : Lin.
927.—Herodias garzetta : Lin.
929.—Bubulcus coromandus : Bodd.
931.—Butorides javanica : Horsf.
937.—Nycticorax griseus : Lin.

H. EDWIN BARNES.

ON THE USES OF PANDANUS OR SCREW PALM,

*Taken from the Journals of the late Handley Sterndale,
with prefatory Remarks,*

By his Brother R. A. STERNDALE, F.R.G.S., F.Z.S.,

Read before the Society on the 7th of December 1885 on
production of specimens of the fruit by Mr. FRAMJEE
N. DAVER.

The *Keora* or *Pandanus Odoratissimus* grows freely throughout India;
whether this is identical with the *Pandanus* of the South Seas, I am
unable to state, but it must be, from my brother's description, of a closely
allied species, and capable of utilization in the same degree. It is, however,
but little known in India for economic purposes, its sole recommendation
being its extremely fragrant flowers, which are used occasionally by
native ladies for adorning their hair.

Roxburgh states that the lower yellow pulpy part of the drupes is sometimes eaten by the natives during times of famine, as also the tender-white base of the leaves, either raw or boiled ; the roots are used by basket-makers to tie their work with, and he adds that they are also used for corks. Small indeed are these results as compared with the manifold purposes to which the tree is put by the South Sea islander. Roxburgh notices that the leaves are composed of longitudinal, tough, and useful fibres like those of the pine apple. Yet this economical product has hitherto been neglected, though the tree is so common in parts that hedges are made of it. In the Nicobar Islands it is called the *Mellore* or bread-fruit, being probably used there for food as it is in the South Pacific. In the Mauritius it is extensively employed in the manufacture of sugar and coffee bags and for export. " Hedge-rows or avenues are formed of it round plantations, or along the sides of the many roads which intersect them, and the leaves, as fast as they attain maturity, are cut till the tree arrives at its full growth, when the production of new leaves being slower and less useful, younger plants are resorted to." So wrote Colonel Hardwicke in 1811. Forbes Royle gives but little information beyond quoting Roxburgh and Hardwicke, and the plant in India has not received much attention. Voight says that in China and Cochin elephants are fed on it. Mr. Stonehewer Cooper, in his " Coral Lands of the Pacific," gives an account of the *Pandanus*, which is evidently taken from my brother's writings, the similarity of expression proving this ; he has acknowledged much of his information so gathered, but might have done more in that way ; however, he has added nothing more to our knowledge of the plant than what will be gained in the following paper, written years before Mr. Cooper's book was published, beyond calling it in one place *Pandanus utilis*, which, according to Voight, is a synonym of *P. odoratissimus*; and stating in another that he does not know of anything that will approach the leaves of the *Pandanus* tree as a paper-making material.* This is a point worth experimenting on, and it is with a view to bring the many qualities of this plant before the public in India, and interest men in what has been hitherto neglected as a jungly thing of no value that I have extracted from my brother's papers, which I hope to publish some day *in extenso*, the following notes on a worthy rival of the Bamboo and the Cocoanut.

" Among the most ubiquitous of vegetable products throughout the Pacific is the *Pandanus* or *Screw Palm*. It is called '*Fara*' in most native tongues, and would seem to a stranger to be as ugly and prickly as it is

---

* I find that Mr. Cooper's account of the *Pandanus*, as well as the remark about its being a good material for the manufacture of paper, is taken *verbatim* without acknowledgment from my brother's report to the New Zealand Government on the Islands of the South Pacific.

densely prolific and apparently useless, but it would be a great error to
suppose so, for it is one of the greatest blessings which Providence has
bestowed upon man in the savage state. It grows, I have heard, upon all the
tropical coasts of Australia, where it is regarded as of no use even by the
Aborigines, but to the savages of the Coral Seas it is food, clothing, shelter,
and an infinity of benefit. It delights in rocky and gravelly soils, impreg-
nated by the salt spray of the sea (or rather where there is no soil, but
gravel only,) and so luxuriates desert isles, where it creates impenetrable
thickets. Its appearance is very singular; when young it looks like a
tussock of 'sword grass,' the edges of the leaves and the ridge in the
middle being fringed with small sharp thorns; these leaves follow each
other spirally up the stalk, so that the tree grows with a perfect twist like
that of a screw auger. In its earlier stages, when about ten or twelve feet
high, it has sometimes a graceful appearance ; as it grows older, it becomes
grotesque; as it is an inhabitant of stony ridges where roots are unable to
penetrate to any depth, and of open coasts exposed to the most furious
winds, it secures itself a hold upon the earth by throwing out around its
butt a number of stays or shrouds, straight, tough and sappy, each of
about the thickness of a man's wrist; they grow round the bole of the tree,
following its spiral formation, and appear first as a sort of wart or excrescence ; this soon takes the form of a horn growing downwards ; it is of a
delicate pink, smooth and glossy, and cuts soft like a cabbage-stalk,
being full of oily sap, which it is important to know will support
life of man or animals where there is no water. It continues to grow
thus until the point touches the ground, where it takes firm root by sending
out a multitude of fibres which penetrate the sand or crevices of rocks,
and wrap themselves securely about the stones. Thus, the brave *Pandanus*
will bend to the hurricane, but start—no, not an inch ! When full grown, it
reaches 30 or 40 feet, and by that time has sent out many odd-looking
limbs branching out from the stem something after the fashion of the
golden candlestick in the Tabernacle of Aaron, each crowned at the end by
a tuft of drooping leaves, a blossom of a pale yellow (something like the
flower of Indian corn and of a strong smell, and a large fruit bigger than
a man's head, outwardly of a dark green colour and in shape resembling
a pine-cone, or the thyrsus represented in the ceremonies of Bacchus.
The trunk of the tree is hollow from end to end, and would make excellent
drain pipes ; the wood is hard as horn and like horn in appearance. I have
seen it when used as pillars in some native houses, scraped and polished as
bright as mahogany. In the ground it soon decays. The fruit consists of
a number of truncated conical polygons, each about 4 inches long,
separate from the others, closely wedged together and radiating from the
interior stalk. The outer ends of these sections are dark green, impenetrably
hard and tough, enclosing eight or ten seeds each, the inner portion, which

is in some species scarlet, in others yellow, has a highly polished surface, and powerful smell like that of a mango; it consists of fibrous pulp in consistence exactly like the interior of a sugar-cane and containing even a larger proportion of saccharine matter; it can be chewed or cut with a knife, and when steamed in an oven seems to consist chiefly of syrup. An intoxicating drink can be made from it by fermenting a mash made of the cooked fruit, as also strong spirits by distillation. The seeds are about the size of a haricot bean and are in appearance and flavour like the kernel of a filbert, so excellent to eat that, were they known, they would be in demand in civilized lands as an article of dessert. But their existence (or nature) is unknown to most Europeans well acquainted with the tree, for as much as these kernels are so concealed and protected as to be almost impossible to get at by those unacquainted with the process. The hard capsules which contain them require to be broken in a peculiar manner by a powerful blow from a heavy stone or sledge hammer, whereby their extraction is very easy. They are wholesome and nutritious. I have on some desert places eaten of them at a time as much as would fill a pint measure. The Polynesians are fond of this fruit, and are constantly chewing the cones; they also thread them on strings after the fashion of a ponderous necklace, so as to form a very gaudy and odoriferous ornament which they eat when they are weary of wearing. Mixed with scraped cocoanut and baked, it is much used on many islands, but as a preserved article of food it is most important, and is in that form peculiar to the Isles of the Equator and the North Pacific. Pounded and dried and packed firmly pressed in baskets, it presents an appearance like coarse saw-dust, and will keep for any length of time. It is called "*Kabobo*," and is the staple article of consumption in many of the equatorial isles and in the Ralik and Ratak chains. Many atolls in these latitudes are destitute of cocoanut trees, so the "*screw palm*" is the sole vegetable subsistence of the inhabitants. The "*Kabobo*" also constitutes the sea stock with which the savage mariners of the Pintados provision their canoes. When required to be eaten, it is mixed with a little water and parched in the sun or baked on hot stones. If it be true that the *Pandanus* grows all round the coasts of North Australia, as I have been assured by seamen that it does, and that the Aborigines of those parts are unacquainted with its use —then do they starve in the midst of plenty—as Solomon says "for lack of knowledge people perish." This I do well know from my own experience that the wastes of very much of New Holland (except where there is absolutely no water either in pools or in '*Mallee*' roots) contain infinitely more means of subsistence for man than such isles as Erikub or Gaspar Rico and other desert cays upon which it has been my fortune to sojourn. But inestimable as is the *Pandanus* in providing food to the inhabitants of desert isles, it is no less valuable to them as the source from whence they derive their shelter, their clothing, and whatsoever

approach to domestic comfort they possess. Their houses are entirely constructed of its timber ; the posts and sills are of the straight columnar
trunk, which are set upright round the whole building about 4 feet apart ;
down each side of the post, in the line of the wall, is cut a groove about
an inch deep, and into these are filled laths which are split with a
knife out of the straight stays which grow round the trunks of these
trees. Thus is made a very neat and comfortable dwelling ; the doors and
window-shutters are made in like manner of the split laths, and the
whole is roofed in with the leaves of the same tree. The thatch is made
very ingeniously : the frame of the roof being complete, a great number
of laths, a fathom long, are split and across them side by side ; the long
leaves are doubled and pinned with thin skewers ; these are laid across
the rafters one over the other and secured with string ; a roof of this
kind looks very neat inside, is impervious to the heaviest rains, and lasts
usually from 10 to 12 years. The floors are made of smooth water-worn
snow-white coral pebbles from the sea beach, which harbour no insects, and
above them are spread mats of this same palm leaf in a double layer,
the lower ones of a coarse make, the upper of a finer kind, so
delightfully cool and smooth that one may lie upon them with great
comfort, absolutely without any clothing between them and the body ;
on some islands they are made very handsome, being of a bright straw
colour, with a stripe four inches wide along each edge and two others down
the middle. This stripe is worked in a variegated pattern in red, yellow
and black ; these colours are obtained by dyes made from the juice of
certain roots. The floor mats are frequently of great size, sometimes
as large as the whole floor, made purposely of corresponding dimensions.
On islands where they make them and sell them to trading ships they
receive payment at the rate of 2 yards of calico for 2 yards square of fine
mat. On islands where the *tapp* tree does not grow, *Pandanus* mats are the
only bed clothes, as also clothing for the body. They consist of soft
ornamented girdles about 9 inches wide and from 12 to 20 feet long,
aprons, pouches and "tiputas ;" these are made very soft and are
bleached between salt-water and sunshine until perfectly white ; the patterns
which are worked into them are also very handsome. The hats which
they make on many isles out of this material are plaited all in one piece,
like those which are made in Guayaquil, and are very neat and durable.
Some baskets (worked in the same manner as the cigar cases so common
in the East Indian islands) they make so very handsome that I have seen
one of them sold for five dollars and counted cheap. On Samoa the
women wear soft *Pandanus* mat for petticoats and trains, which sweep the
ground behind them as they walk on state occasions ; these mats are
generally not handsome, being without ornament except sometimes a little
red fringe, and are of a dirty straw colour ; nevertheless they are consi-

dered so valuable by them that they will sometimes refuse a hundred dollars for one, and would certainly not give it you in exchange for a Cashmere shawl; some of these mats are a hundred years old or more, and full of holes, which does not deteriorate from their value. At a *Samoan* marriage an old mat, which is laid under the bride, is often the most precious article in her whole "trousseau," and has been probably a portion of the dowry of her mother and grandmother. The mat which a fighting chief will sometimes wear about his body is accepted as the ransom of his life if he fall into the hands of his foes. The fortunate victor probably knows the history of it before it comes into his possession, and can tell its age, and where and by whose hands it was woven; the value which they place upon them is wholly fictitious. It is a love of ancient usage which has consecrated them, as the Samoan mats are of mean appearance, and neither so becoming nor so comfortable to wear as two fathoms of cotton print which they might buy for a dollar. The work of making mats and other manufactures from the *Pandanus* leaf is all performed by women. The leaf itself is like that of a flag, two or three inches wide; when gathered, it is laid in the sun to dry; it is then stretched to prevent its curling and to strip it of its thorny edges. For this purpose the women always keep one of their thumb-nails long, as likewise to split the leaf for finer work; such portions as are intended to produce the ornamental part of the pattern are then dyed; the plaiting is performed upon a smooth board with a convex upper surface; as they use their teeth very much in dividing the leaf, they protect their lower lip by wearing upon it the scale of a fish. The time occupied in this work varies according to its texture of the coarser kinds. A woman will plait in a day a yard deep by two yards wide. The sails of canoes on all these islands are made of such mat. The beautifully variegated aprons of the women of Micronesia, and wrappers which the men wear about their loins, consume much time in making; the texture of the fabric being about equal to that of No. 1 canvas, but much softer after being bleached and worn some time. On the low coral isles the finest mats are made, and with wooden dishes, carved pillows, fish-hooks of pearl or turtle shell, lines of cocoanut fibre and 'Ranan' bark are the principal articles of exchange. The 'Ranan' lines are beautiful; they are immensely strong, white as linen, and, though laid up by hand, are equal in regularity of twist and thickness to the best machine-made whipcord or Calcutta white line. These lines are from the dimensions of a packthread to that of a logline which will hold the largest fish; they last *a great number of years*; the savages are very careful of them, washing them with fresh water before putting them away whenever they return from fishing; their finer nets are made of the same bark, which is that of a small tree indigenous to most low coral isles. The making of lines and nets is the work of men. On the Samoan isles, when the necessaries of life were easily obtainable, articles of luxury were in demand, such as fine mats, printed *tappa*, carved and ornamented work, feathers of splendid colours, and oval plates or studs of nautilus shell for the adornment of head-dresses, as also for various purposes hawk-bill, turtle and pearl shell. Besides mats another description of clothing is made by savages from the ' *Pandanus.*' I have mentioned that it throws out stays from the trunk ; these commonly cease to grow out higher than about six feet from the ground, as by that time the growth of the tree upwards has stopped ; before touching the ground, where they take root, their consistence is flexible and sappy. If cut off at this stage and soaked in water after being beaten with a mallet, these stakes are found to consist entirely of fibres agglutinated together by an oily sap ; they are, when well cleaned, pure white, soft and strong like 'jute' or hemp, and are easily obtainable a yard long ; of this fibre they make 'jupons' and a sort of pouches, which are comfortable and serviceable. I have no doubt that this product, if generally known (which it is not), could be turned to some valuable account ; it could be obtained in immense quantity and at no cost but the work of cutting and cleaning, as the *Pandanus* completely overruns many coral islands and desert coasts. When we come to consider the numerous wants of man—food, drink, clothing, shelter and an infinity of comforts—which are supplied by the wood, leaves, fruit, and sap of this remarkable tree ; when we reflect upon the fact that no human being possessing a modicum of ingenuity and the instinct of self-preservation can positively starve where it grows, and that its natural locality is the most desert coasts of the tropic seas, luxuriating, as it does, upon the barren beach immediately contiguous to high-water mark, where there is no soil whatever or apparent moisture ; its nourishment being derived from the arid sand, coral, gravel or boulders of rock, heated throughout the day to a temperature sufficient to burn the human skin, one cannot fail to experience a feeling of astonishment at so striking an evidence of the providence of God."

H. B. STERNDALE.

A NOTE ON PANDANUS ODORATISSIMUS OR SCREW PALM.

(Written at the request of R. A. STERNDALE, Esq., F.Z.S., to follow his paper.)

The *Pandanus* we see here is of two kinds. The yellow variety is generally called *Ketaki* (केतकी) or *Suwarna Ketaki* सुवर्ण कतका, as distinguished from the white Kevadâ (केवडो), or (श्वेत केवडा) Sweta Kevadâ. The yellow variety is much more strongly scented, and is more highly prized by the Hindu ladies, who wear it in their hair. Both these contain staminate organs alone, surrounded by spathaceous bracts; and it is these bracts that form the chief attraction for ladies. The staminate organs of the yellow variety are simpler than those of the white. The anthers of the former are longitudinal, and open longitudinally, giving vent to a fine impalpable powder strongly scented, and forming the pollen. The stamens are almost sessile or about a line in length. These stamens are innumerably crowded in the shape of a cone on a flesh spike or stalk. The anthers of the white variety are shorter and open longitudinally, but their flesh spike is branched. It gives rise to similar impalpable powder, which is gritty, but perhaps less scented, though sufficiently attractive. These clustered and branched staminate spikes go under the name of Kuyali (कुयली), and if they don't get decomposed or rotten during the process of drying, are of great value in keeping off moths from woollen clothes. At least such is their reputation. The stays or aerial roots Mr. Sterndale mentions in his very valuable paper are used in this country by *goundis* or whitewashers for making brushes to whitewash or colourwash houses. The fibrous tissue is separated from the tender interfibrous substance by beating the top of perhaps half a yard bit of the stay or aerial root and made soft and pliable. It makes a capital brush. There is no other use made of *Kevadâ* that I know of. The female flower or collection of flowers turning into fruit is seldom used for any special purpose in this country.

K. R. KIRTIKAR.

ZOOLOGICAL NOTES.

ON VARIATION IN COLOUR IN URSUS LABIATUS, THE SLOTH BEAR, &c.

By R. A. Sterndale.

A CORRESPONDENT in the *Asian* of last week (9th March 1886) gives an account of his killing a sloth bear with two cubs, one of which was brown instead of the usual jet black. In November 1884, I received a letter from Dr. Tomes, Civil Surgeon of Midnapore, asking for my opinion on a skin and skull of a large bear shot in the Midnapore jungle. He described the skin as " a particularly good one, thick and shaggy about the shoulders, of a tawny brown colour throughout, lighter underneath, no black in it anywhere, a whitish collar on chest." Fortunately the skull was preserved, and the dental formula given by Dr. Tomes enabled me to pronounce it an albino of *ursus labiatus*, and not a stray specimen of *ursus isabellinus* escaped from captivity. The sloth bear has, as a rule, two upper incisors less than other bears.

To-day, whilst looking up some correspondence in the *Asian* on another subject, I came across two letters regarding grey bears in the plains of India : one was seen by " H. D. K." writing from Secunderabad, Deccan, of which the hind quarters only were grey; the other was reported by " W. M. R." as seen on the borders of the Shahabad and Mirzapore districts. He says the greater portions of the bear's body was grey, and a light grey too. The Native Shikaris called it a *sufaid bhal.* Unfortunately neither of these two bears was secured. We have, however, ample proof of albinoism in "M.'s" living cub and the Midnapore skin.

ON THE FLYING SQUIRREL OF WESTERN INDIA.

There is no doubt the flying squirrel of this Presidency is *Pteromys Oral,* but the prevailing colour is grey, whereas *Pteromys Oral* is a dusky maroon black grizzled with white. I am inclined to think that it is the same as *Pteromys cinceraceus,* which is in all essential points identical with *Pteromys Oral.* The Society lately received two living specimens from General Watson, which were made over to me for examination. During the night they managed to gnaw a hole through their cage, and escaped. One was re-captured, but the other, I regret to say, has disappeared. Fortunately we retain the finer specimen. Wonderful stories are told concerning the flight of these animals, though flight is a misnomer. They cannot fly as birds and bats do: they merely spring from a considerable height, and the extended skin between their limbs acts as a parachute and floats them along, letting them down easily. Thus they can skim over a space of 50 to 60 yards. Early on the morning of their escape one was observed sitting on a cornice near a window at the northern corner of the Currency Office, where I live. On a servant trying to catch it sprang off in the direction of the Bombay Club and alighted near the Club-door. The distance was sixty-nine paces. These animals are quite nocturnal in their habits, sleeping all day rolled up in a ball with the head tucked in between their fore legs and the tail coiled round the body. At night they are very active.

ON A SPECIES OF PIGMY SHREW.

I would call the attention of Naturalists to the existence of a Pigmy Shrew in the low lands of this Presidency, as more specimens are wanted, and it is possible that on such being found, they are thrown aside under the impression that they are the young of the ordinary species of Musk-rat. The pigmy Shrews are a dwarf race, generally found in the hilly parts of India, Ceylon and Burmah, and they vary in size from 1½ to nearly two inches, exclusive of tail, which is about another inch. They are true Shrews with all the characteristics of the genus, and a Burmese species, *Sorex nudipes,* has the musk glands strongly developed. The Society has received one lately from Mr. Littledale, which he found at Baroda swimming about in a flower-pot during the rains of 1884. He writes: "I kept it alive 3 or 4 days, giving it crickets and flies. It liked to get under a bit of cotton wool, in the shade, and used to make a sudden dash at the cricket if it came near, crunching its back and hind legs first. It has not shrunk at all. It was mouse colour, and the snout pale fleshy. The eyes seemed greyish blue." I have been unable to determine the species as yet; the nearest approach to it is *Sorex perroteti* from the Nilgherries, but it does not agree in colour. *S. perroteti* being blackish brown, whereas this is a pale mouse colour, rather silvery when taken out of the spirit and dried.

R. A. STERNDALE.

ON THE FREQUENCY OF ALBINOISM IN CUTCH, &c.

By Mr. A. T. H. Newnham, S.C., 10th N. I., with

Notes by Mr. E. H. Aitken.

FREQUENT OCCURRENCE OF ALBINOISM IN CUTCH.—Within the last few months the following cases of albinoism have come under my notice, which, I think, are sufficiently numerous to be worthy of mention, viz., May 24, *Chattorrhœa striata* (the striated babbler, presented to Society's collection); July, *Perdicula asiatica*, bush quail (presented to Society's collection) ; and *Molpastes haemorrhous*, the Madras Bulbul, partially so, the wings only being white. It would seem as though the prevailing tint of the country, which is principally composed of sandy plains, had some influence on the colouring of its inhabitants, for the birds generally are of a paler colour than the same species which I have noticed where the soil is darker in tone. Besides these specimens which have been shot and obtained, there have been other occurrences. Last year a perfectly white *Sarkidiornis melanonotus*, more generally known among sportsmen as the Nukta, used frequently to be seen on one of the sacred tanks of Bhuj, where unfortunately it is forbidden to shoot, and a second case of *P. Asiatica* was met with out shooting. The latter, strictly speaking, was of a soft fawn colour rather than white. A white squirrel also used to haunt one of the bungalows here.

THE BHALU.—We are occasionally visited at night by one of these mysterious Janwars. There are various explanations given as to what it really is ; some asserting that it is a lynx, others a female jackal, and others that it is an old worn-out jackal, which follows in the tracks of some larger animal to obtain its leavings. I know the latter is the more general belief, but though I have made frequent enquiries from the Shikaris and villagers here I have not come across any one yet who has actually seen one. The cry is a sort of convulsive scream ending abruptly in a hoarse crack. I never hear it at night without sallying forth with a gun to try and shoot it and clear up the mystery for myself, but hitherto without success. One moonlight night I heard its cry quite close to me, but could distinguish nothing. Perhaps some of your correspondents can enlighten me as to what it is.*

A. T. H. NEWNHAM.

NOTE BY MR. AITKEN.—Mr. Newnham's observations are supported by several things that came under my notice during a year's residence at Kharaghora on the borders of the Runn of Cutch. There were not many species of butterflies at the place, but the two commonest, *Danais chrysippus* and *Papili siphilus*, were often conspicuously pale and colourless. They would have been considered *poor* specimens if caught in Bombay. I believe that variety of the former, with a dash of white on the hind wings, which has been separated under the name of *D. alcippoides*, is only a stronger exemplification of the same effect. It would probably be found to be not uncommon in this region.

* The Kol Bhalu, Pheal, Pheeou, Phinkarr, or Sial, is an ordinary jackal. Several have been shot in the act of howling, and there was nothing abnormal about them. The subject was well ventilated in the *Asian* in 1881-82, and the general opinion pointed to the above conclusion. Correspondents gave evidence from all parts of India.—R, A. S.

Of four specimens of *D. dorippus* from Aden, now in the Society's collection, two exhibit this dash of white. On the other hand, collections of butterflies caught among the luxurious vegetation of Khandalla or Matheran generally contain specimens with a depth of colour never met with on the plains.

But the strangest instance of the effect of an arid, sandy country on animal colour, if it was really an instance, was a mungoose which I repeatedly saw at Kharaghora, but did not secure. It was apparently the common mungoo e of Bombay (*H. Griseus*),* but the tip of its tail, instead of being blackish, was white. A solitary "sport" like this has not much significance by itself, but it becomes suggestive when we remember that the desert fox of Cutch (*leucopus*) differs from the common Indian fox in this very point that its tail is tipped with white instead of black.   E. H. A.

BOTANICAL NOTES.

ON AN INSTANCE OF FRUCTIFICATION IN A STAMINIFEROUS PLANT, CARICA PAPAYA.

By Surgeon-Major G. Bainbridge, I.M.D.

The *Papayaceæ* form a small order of three or four genera and 25 or 30 species only, not very distantly related to the cucumbers. The species are all tropical, and several inhabit S. America, of which the plant under notice is supposed to be a native.

Carica Papaya is the best-known individual of its order, and has excited much interest owing to the presence in its tissues of *Papain*, an alkaloid or principle having the property of digesting animal substances, and serviceable, therefore, as a medicinal agent.

As is well known, the plant is normally diœcious and one of the most conspicuous examples of this marital arrangement. You will all have distinguished the male, with its long-stalked panicles of small yellowish flowers, from the female or pistilliferous tree, with its much larger, whitish, rather campanalate flowers, which are closely arranged around the trunk and branches, under the shelter of the leaves, and, having very short stalks, are nearly sessile.

I was not aware until recently that this arrangement was ever departed from. But in January last year (1884) I was surprised to find at Dharwar, in the garden of a house I had just entered, a male *Papaya* tree bearing fruit upon its long pendent stalks.

On examination I found its flowers to resemble the typical male ones in every respect, except in the presence of a minute ovary in at least some of them.

By April the fruit had grown to a considerable size, so that some of them measured ten and thirteen inches in circumference; and, what was more interesting, they contained numbers of ripe black seeds about three-fourths

* Probably *H. Ferrugineus*, Sind species, the tail of which is lighter coloured normally.— R. A. S.

q

o
a
g
is
pe

to th

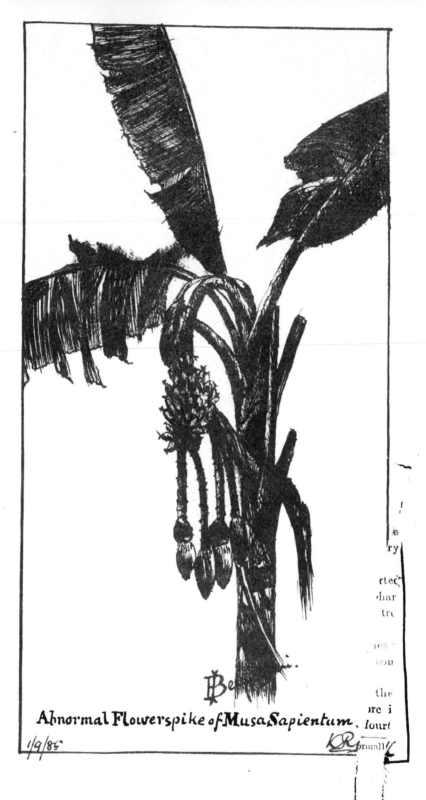

Abnormal Flowerspike of Musa Sapientum.

1/9/85

of the size of normal ones from well-grown fruit. I saved a large number of them, intending to try whether they would germinate; but they were lost in the hurry of my transfer.

I now show three small specimens of the fruit of the same tree which I have had sent to me. The largest measures six inches in circumference· I also present a rough sketch of the tree drawn in April last.

I imagined this curious occurrence to be almost unique ; for its possibility was hitherto unknown to me, though I have seen much of *Papaya* cultivation for some years. I find, however, that the fact is noted by Roxburgh in the Flora Indica, 1832, where he mentions two instances, and states that the same is common at Malacca.

Botanical class books and other authorities which I have examined do not mention the matter. There seem to be two varieties of *Papaya*, one producing rather globular, and the other citron-shaped, fruit of much larger size and superior quality. Fertility of soil may, however, possibly account for this and for the strange " variation " to which I have drawn attention.

G. BAINBRIDGE.

ON ABNORMAL DEVELOPMENT IN MUSA SAPIENTUM.
By Surgeon K. R. Kirtikar, I.M.D.

Read on 1st September 1885 before the Botanical Section.

I submit a photographic print* of an abnormal development of the flowerstalk of *Musa Sapientum* (Banana) growing in a garden on Girgaum Back Road, Bombay. The drooping spike, after having thrown out two or three clusters of flowers in the axils of the first two or three purple fleshy bracts, sub-divides and thus forms two spikes instead of a single central. The primary spike remains thicker than the secondary stalk, as the division of the spike is not strictly dichotomous. The secondary grows longer and sub-divides again. The primary also, after throwing a few more clusters of flowers, sub-divide again into two spikelets. Thus, there are four spikelets instead of one spike. The final or apical buds, sheathed in their purple bracts, still remain, with a few abortive flowers. I call these "abortive" flowers, because they never turn into the fruit called banana or plantain, but open and die.

NOTE ON AGARICUS OSTREATUS.

The Fungus described by Dr. Dymock in his Vegetable Materia Medica of Western India (p. 704, 1st Edition) is called Phanasamba in Marathi and named by him as *Agaricus ostreatus*. *Agaricus ostreatus* often does grow on jackfruit tree. But on examining genuine specimens of what is usually gathered and sold and used under the name of *Phanasamba*, it appears to be a *Polyporus* and not an *Agaricus*. (See Badham's Esculent

---

* A water-colour drawing from the same has since been presented by Surgeon Kirtikar to the Society, of which a lithographic print accompanies this.

Fungi, Plate X., and Mrs. Hassey's Illustrations of British Mycology, XIX. Plate, Second Series). Dr. Sakharam Arjun, following old descriptions, also calls the fungus *Agaricus ostreatus.*

But a figure of the *Polyporus* is given in Batsch's Elenchus Fungoram, Plate XLI., page 114, Continuatio Secunda. It is called *Boletus* "*Nitens*" or *Crocatus.* It appears a proper description of Phanasamba has not yet appeared. I exhibit several specimens, a general description of which will appear in my work on the Bombay Fungi, which I hope will be published at no distant date. As this variety of *Polyporus* mainly derives its name from its habitat—growing on *Phanas* or Jack tree,—I have named it *Boletus Nitens Artocarpalis.*

ON THE FRUIT OF TRAPA BISPINOSA.

The fruit of *Trapa Bispinosa* (exhibited along with the plant in flower), *Shingâda* as known among the Hindus.

The fruit resembles, roughly speaking, a bullock's head in miniature, and is an important and highly-prized article of diet among the Hindus. The whole of the fruit is mealy, and is as delicious when baked or boiled as a chestnut. Peeled, pounded and boiled with milk and sugar, it forms an excellent repast under the name of *hulwa*, and deserves to be more widely known. It is eaten either fresh, or is peeled and dried for use afterwards. For drying, only the mature fruit is serviceable; if it is not mature, it shrivels up and often decays. Mixed with pepper, salt and cocoanut kernel scrapings and fried in ghee or clarified butter, in lumps as big as a cherry or plum, it is very delicious. It is highly valued by the Guzrathis, and is generally sold dried in a *Kirani's* shop (seller of groceries and spices), and very largely used on fast days, when rice, wheat, and such other daily articles of food are not eaten.

The plant which bears this fruit is an aquatic annual, and grows very quickly. It is cultivated largely in tanks around Thana, the young sprouts being simply deposited on the surface of the water. It flowers about August and September, and fruit is gathered about November. If the old and dead decaying leaves are removed as they form from time to time, the tanks in which the plant is cultivated have clear water, probably from destroying minor vegetable life on which it feeds, or at any rate partially derives its nourishment.

NOTE ON KASRA OR SCIRPUS KYSOOR.

Read on 22nd January 1886.

I exhibit to-day three articles—(1) the boiled hairy root-bulb ; (2) the same boiled and peeled; (3) a *hulwa* made of the peeled bulb. I also exhibit along with specimen No. 3 a *hulwa* made of the fruit of *Shingâda* referred to in my Notes read before this Section at our Septem-

BALANOPHORA Sp.

ber Meeting. Pounds and pounds of this delicious bulb are used as an article of diet on fast days among Hindus. The root bulb is often sold dried after being peeled. The plant itself belongs to the Sedgewort family, and is described at p. 288 of Dalzell and Gibson's Flora. The bulbs are gathered in January, February and March, after the plant dies. I exhibit the plant here. It thrives in the rainy season, and grows abundantly in tanks round Thana. The skin of the bulb is hairy; the rootlets being often two or three inches long and tufted at the apex, or extreme end. The roots sometimes shoot out in rings round the body of the bulb. The leaf of the plant is hispid, 3 to 5 feet lòng, studded with oblong air spaces. The plant flowers in the rainy season about July or August, and having lived its annual life, dies away. It is after this that the bulbs are gathered; they are edible even uncooked, but are not very palatable. They are usually in very great quest, and are obtainable at one anna a hundred bulbs. The *hulwa* made with sugar and milk is considered a dainty. This *hulwa* is more glutinous than the *hulwa* of *Shingâda.* It would be interesting to find out the relative food-value of these important articles of diet, especially as regards the proportion of starch gluten, and salts. The leaf does not seem to be sufficiently strong for any of the purposes for which common bulrushes are used, such as for making mats, baskets, chair bottoms, nor do I know of any medicinal uses of the plant. At page 721 of his Vegetable Materia Medica, Dr. Dymock asks a question as to whether *Kasceroo* (Hind) is the *Scirpus Kysoor* of Roxburgh. I am certain it is. Dr. Dymock also says it is given in diarrhœa and vomiting. If in addition to its value as a delicate article of food, it is really useful in diarrhœa, a congee made of it with milk will be a very suitable form of nourishment in diarrhœa cases and in vomiting. I can bear testimony to its bland and soothing properties. The boiled bulb with common salt is very delicate eating.

K. R. KIRTIKAR.

NOTE ON A SUPPOSED ROOT-PARASITE FOUND AT MAHABLESHWAR IN OCTOBER 1885.

By Mrs. W. E. HART, *Read on 15th March* 1886.

IN October a tuberous-rooted plant of curious structure, which I have endeavoured to sketch below, was brought to me at Mahableshwar, from one of the valley jungles below the hill. The rains had continued more than usually late, which may account for there being then still visible a plant which neither I nor any one to whom I showed it had ever seen before. It grew in clusters in moist red laterite clay, through which occurred the numerous root fibres (lately severed) of some large dicotyledonous tree. The man who brought me the plant declared that he very rarely met with it, never except during the rains, and then only in the

thickest jungle, and always at the foot of some large tree. But he was
unable to state whether the large tree was always of the same species. The
first thing to appear above the soil was a yellow spathaceous stalk, bearing
on its summit a ball, about the size of a marble, almost concealed among
the spathes. Most of these balls were of a velvety texture and a rich brown
colour. Two were rough, not unlike fir-cones. The balls continued to
grow in circumference as the stalks grew in height, till the latter were
about 3 inches long and the balls about the size of bagatelle balls. A
number of minute white flowers then opened over the whole surface of
the rough ball. Having no microscope or magnifying glass with me,
I was unable to identify the plant from the examination of its extremely
minute structural parts. Dr. Macdonald determined the open flowers on
the rough heads to be staminal only, and conjectured the velvety balls to
be composed of pistillate flowers only, and from the stamens being sinuous
and united into a central column, he was inclined to think the plant
might possibly belong to the Natural order Cucurbitaceæ. But as he
also had no magnifier, he was unable to speak with certainty, and failed
to identify the plant. I much doubt if there is any Cucurbitaceous plant
without the climbing habit so characteristic of that order. On the other
hand, the small Natural orders Cytinaceæ and Balanophoraceæ, especially
the latter, present some features similar to those noticed in my Mahableshwar plant. The following characteristics at least of Balanophoraceæ,
as described by Dr. Balfour in his Class Book of Botany, seem to be
identical with both those noticed by himself and those determined by
Dr. Macdonald :—" Leafless...with tubers...whence proceed naked or scaly
peduncles bearing heads of unisexual flowers. Staminal flowers generally
white...anthers...united into a multicellular mass...Parasitic on the roots
of various dicotyledons, and abounding on the mountains of tropical
countries." Dr. Balfour certainly says nothing of the very curious and
characteristic velvety ball, nor was the parasitic nature of my plant fully
established, but what I ascertained of its habits from the man who
brought it to me is at least not inconsistent with its being a root-parasite
He also informed me that the plants died down in the dry weather
and had never been known to survive removal. I kept mine alive
for some weeks in a soup-plate of water, but it was completely withered
before I left Mahableshwar in January.

   A.—Brown velvety ball.

   B.—Rough ball covered with minute white flowers.

   C.—Spathaceous stalk, greenish-yellow towards the top and brighter-yellow in the
      lower part.

   D.—Lump of red marly earth, apparently moist laterite clay, containing numerous
      root fibres, in which the plant was growing.

I incline to the belief that the plant was one of the Balanophoraceæ. But
the man who brought it to me was not aware of its possessing any useful

properties, nor did he know any native name for it, though Dr. Balfour describes the Balanophoraceæ as being some of them styptic, and others edible. I can find no reference to Balanophoraceæ in Dalzell and Gibson's "Bombay Flora," published in 1861, nor in Gell's "Handbook for use in the Jungles of Western India," published in 1863, nor in Drury's "Useful Plants of India," published in 1873. In the "Cyclopædia of Natural History," published by Bradbury and Evans in 1856, two years after Dr. Balfour's "Class Book," the Balanophoraceæ are described as "a natural order of parasitical plants growing upon the roots of woody plants in tropical countries and rooting into wood from which they draw their nutriment. . . . . None of the species have fully formed leaves, but closely packed fleshy scales clothe their stems and guard their flowers in their infancy. Succulent in texture, dingy in colour, and often springing from a brown and shapeless root stock, Balanophoraceæ remind the observer of fungi more than of flowering plants, and in fact they appear intermediate in nature between the two. If they have flowers and sexes both are of the simplest kind, and their ovules, instead of changing to seeds like those of other flowering plants, become, according to Endlicher, bags of spores, like those of true flowerless plants. Even their woody system is of the most imperfect kind, for it is either entirely, or almost entirely, destitute of spiral vessels." This writer also notices the styptic and edible properties of certain species.* Again, however, nothing is said of the large velvety ball, so striking in my specimen. It is figured in the illustration to the article which I have quoted, but as oval in shape, and small in size in proportion to the length of the stalk, which, again, is represented as smooth and slender.

I have trespassed at this length on your patience, because, if I am right in my theory that my plant was a Balanophora, it is interesting to botanists for two reasons: *first*, as being hitherto undescribed in the Flora of this Presidency; and *secondly*, and specially, as being, apparently, a link connecting the fungi directly with the flowering plants, without the intervention of the Ferns and other higher orders of Cryptogams, which may possibly be of value in the discussion of the Darwinian theory of evolution.

J. B. H.

---

* Dr. Dymock, in his "Materia Medica of Western India," states that a drug is sold in Bombay called by the natives *Gaj Pipal*, which Messrs. S Arjun and N. M. Khan Sahib consider to be the entire plant of a Balanophora. It appears to be of a different species to the above, and is considered mucilaginous and astringent.—J. B. H.

MEMORANDUM by Dr. D. Macdonald, m.d., *Vice-President of the Society*, on the Species of BALANO-PHORA, Found and Described by Mrs. W. E. Hart.

THE plant consisted of an irregular, somewhat flattened rhizome, roughly tubercular on its upper surface, and having the remains of rootlets on its under surface. On the upper surface of the rhizome there were several short unbranched cylindrical peduncles, ar. inch or more in diameter, more or less completely covered by imbricated fleshy scales, of a yellowish colour; the peduncle terminated in a rounded convex head, on which were studded numerous flowers. These heads were of two kinds—one being covered with staminate flowers, consisting of a deeply four-lobed perianth, enclosing a central column or androphore, and having the anthers arranged in a sinuous in form on its summit. The second kind of head was soft and velvety to the touch; but the separate flowers, which were densely packed, were too small to admit of identification without a magnifying glass. One or two small portions I tried to preserve, and after leaving the hills I was able to make out that they were pistillate flowers, with a minute ovary, and a simple style and stigma.

My first impression was that the plant was possibly a peculiar cucurbitaceous plant, seeing the flowers were monæcious, and that the staminate flowers had monadelphous stamens with sinuous anthers. But on returning to Bombay I found the characters answered to the descriptions given of the Balanophoraceæ—an order which Hooker has studied with great minuteness, and which has many points of special interest.

More than thirty years ago botanists grouped several orders—Cytinaceæ, Rafflesiaceæ and Balanophoraceæ—into a separate class, which was placed between the flowering and non-flowering plants. These orders had a few characters in common: they were parasitical; destitute of true leaves; the stem was generally an amorphous fungoid mass, and there was an absence of green colour. The nature of the seeds was little known, some being described as consisting of a mass of spores, and others as having a cellular nucleus. The researches of Hooker and others have shown that there were not sufficient grounds for forming a new class, and now these orders are looked upon as simply degraded exogens. Hooker considers the Balanophoraceæ allied to the Natural Order Haloragaceæ. Lindley and others have confirmed the statement of the elder Richard that the seeds of at least some plants of the order contain an embryo, which is minute, globular, and undivided.

The Balanophoraceæ have been likened to fungi from their appearance and mode of growth, but they differ from fungi in consistence, anatomy, structure, slow mode of growth, and in having conspicuous male flowers. The parasitism of the plants is of such a nature that there is some difficulty in making out where the tissues of the host-plant end, and those of the parasite commences, as the vascular tissue of the one is continuous with that of the other.

The Balanophoraceæ are parasitic on the roots of trees, and are found in the mountains of tropical countries. Several species are found in the Himalayas, and in the Khasya Hills, and eight or ten species are stated by Griffiths to inhabit the Indian continent. One plant—a Balanophore—is mentioned in a list of plants in the N. W. as being sold in the bazars under the native name of *Gochamúl ;* and another in Kashmir, or another name for the same plant, *Gargazmúl.* But I am not aware of any Balanophora having been described as found in the Bombay Presidency.

Astringency is common to most plants of the order, and one (*Fungus melitensis*) was known so far back as the time of the Crusades, when it was used medicinally as a styptic. A few of the plants are edible, one of which, known in Peru as Mountain Maize, grows with wonderful rapidity after rain. In this plant it is not the rhizome, but the scape, or flowering stalk, which is used. It is said to be eaten like mushrooms, which it resembles in outward configuration. Candles are made from a hydrocarbon obtained from a Java Balanophor.

The Mahableshwar plant is a Balanophor, and undoubtedly it belongs to the tribe Eubalanophoreæ, as it is the only tribe of the seven into which the order is divided in which the perianth of the staminate flowers is four-lobed, and the stamens monadelphous. It is not impossible that Mrs. Hart's paper may be the means of drawing attention to any monograph or publication in which the plant is described, if any exists. Should any member of the B. N. H. S. be fortunate enough to produce another specimen of the same plant, it would probably be best preserved in spirit.

D. M.

LIST OF BIRD SKINS FROM THE
SOUTH KONKAN.

(*Ratnagiri and Savantvadi*)

Presented to the Society by Mr. G. W. VIDAL, C.S., January 1886.

No. in List of Birds of India.	Species.	No. and Sex of Specimens.	Total No. of Skins of each species.
2	Otogyps calvus—Scop.................	m	1
8	Falco perigrinus—Gm.	m	1
17	Cerchneis tinnunculus—Lin.	m m m m i	5
23	Astur badius—Gm.	m f f	3
31	Hieraetus pennatus—Gm.	m m	2
35	Limnaetus cirrhatus—Gm.	f	1
39 *bis*	Spilornis melanotis—Jerd.	i	1
48	Butastur teesa	f	1
51	Circus macrurus—S. G. Gm.	m f f	3
54	Circus aeruginosus—Lin.	m	1
55	Haliastur indus—Bodd.............	m	1
56	Milvus govinda—Sykes	m	1
60	Strix javanica—Gm..................	m	1
65	Syrnium ocellatum—Less.	m f	2
74 *sept*	Scops brucii—Hume.................	f	1
75 *guat*	Scops malabaricus—Jerd............	i i	2
76	Carine brama—Tem.	f f f f f i	6
78	Glaucidium malabaricum—Bly. [not typical, but intermediate between *malabaricum* and *radiatum* (77)]	m f i	3
82	Hirundo rustica—Lin	m	1
84	Hirundo filifera—Steph.	f f	2
90	Ptyonoprogne concolor.............	i i	2
102	Cypsellus batassiensis—J. E. Gr...	i	1
103	Collocalia unicolor—Jerd............	m m m m m f i i i i i	11
107	Caprimulgus indicus—Lath.	m f f	3
112	—————— asiaticus—Lath.	m m m m	4
114	—————— monticolus—Frankl...	m m f f	4
117	Merops viridis—Lin.	i i	2
118	—————— philippinus—Lin............	m f	2
123	Coracias indica—Lin	f	1
127	Pelargopsis gurial—Pears	i	1
129	Halcyon Smyrnensis—Lin	m m m f f	5
132	—————chloris—Bodd	m i	2
134	Alcedo bengalensis—Gm.	i	1
136	Ceryle rudis—Lin.	m	1
140	Dichoceros cavatus—Shaw	f	1
141	Hydrocissa coronata—Bodd.	m f f f	4
148	Palaeornis torquatus—Bodd	m m f	3
149	—————purpureus—P.L.S. Mull	m	1
151	————— columboides—Vig......	m f f i	4
153	Loriculus vernalis—Sparrm.	m m f f	4
160	Picus mahrattensis—Lath.	m m f	3
164	Yungipicus nanus—Vig.............	m	1
167	Chrysocolaptes festivus—Bodd....	m f f	3
179	Micropternus gularis—Jerd. [not typical, almost as near *phaeoceps* (178)]	m f f	3
181	Brachypternus puncticollis—Malh.	m m f f f f f f f f	10
193 *bis*	Megalaema inornata—Wald.........	m m f f f	5
194	—————— viridis—Bodd	m m	2

No. in List of Birds of India.	Species.	No. and Sex of Specimens.	Total No. of Skins of each species
197	Xantholœma hæmacephala— P. L. S. Müll....................	*m m m f*	4
202	Cuculus sonnerati—Lath	*m*	1
205	Hierococcyx varius—Vahl.	*i*	1
208	Cacomantis passerinus—Vahl......	*f*	1
212	Coccystes jacobinus—Bodd	*m*	1
213	———— coromandus—Lin	*m*	1
214	Eudynamis honorata—Lin	*m m f f f f*	6
217	Centrococcyx rufipennis—Ill	*m f i i i*	5
219	Taccocua leschenaulti—Less	*m m*	2
226	Æthopyga vigorsi—Sykes	*m*	1
232	Cinnyris zeylonica—Lin	*f i*	2
234	———— asiatica—Lath	*m i*	2
235	——- lotenia—Lin	*m m*	2
238	Dicæum erythrorhynchus—Lath..	*i i*	2
239	———— concolor—Jerd.	*i*	1
240	Piprisoma agile—Tich.	*f f*	2
254	Upupa epops—Lin.	*f f*	2
257	Lanius erythronotus—Vig.	*f*	1
260	———— vittatus—Valenc............	*f*	1
265	Tephrodornis pondicerianus—Gm.	*m m*	2
267	Hemipus picatus—Sykes	*m f*	2
268	Volvocivora Sykesi—Strickl.	*m m f f*	4
270	Graucalus macii—Less.............	*m m f i*	4
272	Pericrocotus flammeus—Forst ...	*m f*	2
276	———— perigrinus—Lin	*m*	1
278	Buchanga atra—Herm.............	*m f*	2
280	——- longicaudata—Hay	*m*	1
281	——- cærulescens—Lin.	*m*	1
285	Dissemurus paradiseus—Lin.	*m m f f*	4
286	Chibia hottentota—Lin.	*f f f f*	4
287	Artamus fuscus—Vieill	*m m f*	3
288	Muscipeta paradisi—Lin	*m m m m*	4
290	Hypothymis azurea—Bodd.........	*m m f f*	4
293	Leucocerca leucogaster—Cuv.....	*m m i*	3
297	Alseonax latirostris—Raffl	*m m*	2
301	Stoporala melanops—Vig.	*m f i*	3
306	Cyornis tickelli—Bly.	*m m*	2
342	Myiophoneus Horsfieldi—Vig......	*m f i i*	4
345	Pitta brachyura—Lin	*m m m m*	4
351	Cyanocinclus cyanus—Lin.........	*m f*	2
353	Petrophila cinclorhyncha—Vig. ...	*m m m m f f*	6
354	Geocichla cyanotes—Jard. & Selb.	*m m f f i*	5
355	———— citrina—Lath	*m*	1
359	Merula nigropilea—Lafr.............	*m m m f f f f f*	8
385	Pyctoris sinensis—Gm.	*f f f*	3
389	Alcippe poiocephala—Jerd..........	*m f i*	3
398	Dumetia albogularis—Bly.	*i i*	2
399	Pellorneum ruficeps—Sws.	*f f f f*	4
404	Pomatorhinus Horsfieldi—Sykes...	*m m m f*	4
435	Malacocercus Somervillii—Sykes.	*m m m m m f f f f i i i i*	13
446	Hypsipetes ganesa—Sykes	*f*	1
450	Criniger ictericus—Strickl.	*f f i i i*	5
452	Ixus luteolus—Less.................	*m f f*	3
460 *bis*	Otocompsa fuscicaudata—Gould.	*m m*	2
462	Molpastes bæmorrhous—Gm.	*f*	1
463	Phyllornis Jerdoni	*m m m m f*	5
468	Iora tiphia—Lin.	*m m i i*	4
470	Oriolus kundoo—Sykes	*m m*	2
472	———— melanocephalus—Lin. ...	*m m f i*	4
475	Copsychus saularis—Lin.............	*m*	1

No. in List of Birds of India.	Species.	No. and Sex of Specimens.	Total No. of Skins of each species.
476	Cercotrichas macrura—Gm.	*m*	1
479	Thamnobia fulicata—Lin.	*f f*	2
481	Pratincola caprata—Lin.	*m f*	2
483	———— indicus—Bly.............	*i*	1
497	Ruticilla rufiventris—Vieill.........	*m m*	2
514	Cyanecula suecica—Lin.	*m f*	2
515	Acrocephalus stentorius—Hemp. and Ehr...	*m m f*	3
516	———— dumetorum—Bly...	*m*	1
534	Prinia socialis—Sykes	*i*	1
538	——— Hodgsoni—Bly..............	*i i*	2
543	Drymœca inornata—Sykes	*i*	1
544 *bis*	———— rufescens—Hume.	*m m m f*	4
559	Phylloscopus nitidus—Bly	*m f*	2
560	———— viridanus—Bly	*m i i i*	4
563	Reguloides occipitalis—Jerd	*m*	1
589	Motacilla maderaspatensis—Gm...	*m m f*	3
591 *bis*	————dukhunensis—Sykes......	*i*	1
592	Calobates melanope—Pall.	*m*	1
593	Budytes cinereocapilla—Savi	*m*	1
595	Limonidromus indicus—Gm.......	*m*	1
597	Anthus trivialis—Lin.	*m i*	2
600	Corydalla rufula—Vieill	*m i*	2
631	Zosterops palpebrosa—Tem.......	*m i*	2
648	Machlolophus aplonotus—Bly.......	*m f*	2
660	Corvus macrorhynchus—Wagl. ...	*m*	1
663	———— splendens—Vieill	*m*	1
674	Dendrocitta rufa—Scop.	*m f f*	3
684	Acridotheres tristes—Lin.	*m f*	2
686	———— fuscus—Wagl.	*m m m f f f*	6
687	Sturnia pagodarum—Gm.	*m m m f i*	5
688	———— malabarica—Gm.	*m m m f f f*	6
690	Pastor roseus.................	*m m m*	3
698	Amadina rubronigra—Hodg.	*i*	1
699	———— punctulata..................	*m f f*	3
706	Passer domesticus—Lin	*m*	1
711	Gymnoris flavicollis—Frankl	*f*	1
721	Euspiza melanocephala—Scop......	*m*	1
758	Ammomanes phœnicura—Frankl .	*m m f*	3
760	Pyrrhulauda grisea—Scop	*i i*	2
765 *bis*	Spizalauda malabarica—Scop	*m m f*	3
773	Crocopus chlorigaster—Bly........	*m i*	2
775	Osmotreron malabarica—Jerd......	*f*	1
786	Palumbus Elphinstonii—Sykes ...	*f*	1
788	Columba intermedia—Strickl	*f*	1
794	Turtur senegalensis—Lin.	*i*	1
797	———— tranquebaricus—Herm ...	*m*	1
798	Chalcophaps indica—Lin............	*m*	1
803	Pavo cristatus—Lin..................	*m f*	2
814	Galloperdix spadiceus—Gm.	*m f f*	3
826	Perdicula asiatica—Lath............	*m m m m m m f f*	8
829	Coturnix communis—Bonn	*m*	1
830	———— coromandelica—Gm......	*f*	1
832	Turnix taigoor—Sykes	*f f f*	3
840	Cursorius coromandelicus—Gm ...	*m m i*	3
846	Ægialitis geoffroyi—Wagl.........	*f*	1
847	———— mongola—Pall	*m f f*	3
856	Lobipluvia malabarica—Bodd......	*i*	1
859	Œdicnemus scolopax—S. G. Gm.	*m m f*	3
872	Gallinago gallinula—Lin.	*i*	1
873	Rhynchœa bengalensis—Lin.	*m*	1

No. in List of Birds of India.	Species.	No. and Sex of Specimens.	Total No. of Skins of each species
877	Numenius lineatus—Cuv...........	*f*	1
878	——— phæopus—Lin.	*m*	1
882	Tringa subarquata—Güld	*m*	1
884	——— minuta—Leist	*i*	1
893	Tringoides hypoleucus—Lin.	*i*	1
894	Totanus glottis—Lin.	*i*	1
898	Himantopus candidus—Bonn......	*f*	1
901	Hydrophasianus chirurgus—Scop.	*f*	1
903	Fulica atra—Lin.	*m*	1
905	Gallinula chloropus—Lin...........	*m f*	2
907	Erythra phœnicura—Penn	*f*	1
910	Porzana bailloni—Vieill	*m*	1
931	Butorides javanica—Horsf.........	*f*	1
964	Querquedula crecca—Lin...........	*f*	1
971	Fuligula cristata—Lin..............	*f f*	2
978 *tes*	Larus affinis—Reinh...............	*f*	1
980	——— brunneicephalus—Jerd	*m f f f*	4
987 *bis*	Sterna albigena—Licht	*f*	1
Total No. of Species 185.			Total No. of Skins 444

LIST OF BIRD SKINS FROM BURMAH AND OTHER PARTS OF INDIA.

Presented to the Society by Mr. G. W. VIDAL, C.S., January 1886.

No. in List of Birds of India.	Species.	Locality.	No. and Sex.	No. of Skins of each species.
23 *bis*	Astur poliopsis—Hume.	Borongho	*m*	1
39	Spilornis cheela—Lath.........	Akyab	*f*	1
55	Haliastur indus—Bodd	Calcutta.........	*f*	1
74 *sept*	Scops Brucii—Hume	Poona............	*m*	1
77	Glaucidium radiatum (typical)—Tick	Raipur, C.P.	*m*	1
142	Hydrocissa albirostris—Shaw.	Burmah	*f*	1
144	Meniceros bicornis (Ocyceros birostris)	Singbhoom ...	*i*	1
146 *ter*	Rhyticeros subruficollis—Bly	Amherst..........	*i*	1
180	Brachypternus aurantius—Lin	Raipur, C.P. ...	*m*	1
215	Rhopodytes tristis—Less......	China, Baheer..	*i*	1
239	Dicæum concolor—Jerd.	Kotagherry ...	*m*	1
257 *bis*	Lanius caniceps—Bly.	N. Kanara ...	*i*	1
261	——— cristatus—Lin..........	......	*i*	1
306	Cyornis tickelli—Bly.	Saugor, C. P....	*m*	1
360	Merula simillima—Jerd.	Ootacamund ...	*m*	1
434	Malacocercus malabaricus—Jerd.	Ootacamund ...	*m*	1
452 *dec*	Iole viridescens—Bly.	Amherst ...	*f*	1
534	Prinia socialis—Sykes	Madras ...	*f*	1
596	Anthus maculatus—Hodgsn .	Etawah ...	*m*	1
781 *bis*	Carpophaga cupræa—Jerd. ...	N. Kanara ...	*i*	1
843	Glareola lactea—Tem.	Murdan	*i*	1
847	Ægialitis mongola—Pall	Karachi	*m*	1
848	——— cantiana—Lath ...	Diamond island	*m*	1
859	Œdicnemus scolopax—S. G. Gm.	Etawah ...	*i*	1
24				24

CATALOGUE OF SNAKES IN THE SOCIETY'S COLLECTION.

Family.	Genera and Species.	Locality.
I.—TYPHLOPIDÆ (Blind Snakes.)	Typhlops porrectus	Bandora.
II.—TORTRICIDÆ (Short-tailed Earth Snakes)	None.	
III.—PYTHONIDÆ (Pythons.)	Python molurus..............	Lanowli.
	Python reticulatus	In a ship from Rangoon.
IV.—ERYCIDÆ (Sand Snakes.)	Eryx johnii	
	Do.	
	Gongylophis conicus	
	Do.	
V.—ACROCHORDIDÆ (Wart Snakes.)	Chersydrus granulatus......	Bombay Harbour.
	Do.	Do.
	Do.	Do.
	Do.	Alibag.
VI.—UROPELTIDÆ (Rough-tailed Earth Snakes)	Silybura brevis	Khandalla.
VII.—XENOPELTIDÆ (Iridiscent Earth Snakes)	None.	
VIII.—CALAMARIDÆ (Dwarf Snakes.)	Aspidura trachyprocta	Ceylon.
	Do.	Do.
	Do.	Do.
	Do.	Do.
IX.—HOMALOPSIDÆ (River Snakes.)	Cerberus rhynchops	Alibag.
	Do.	Do.
	Do.	Born in Society's rooms.
	Do.	Do.
	Sp. nova....................	Saugor, C. P.
	Do.	
X.--AMBLYCEPHALIDÆ (Blunt-headed Snakes.)	None.	
XI.—OLIGODONTIDÆ (Filleted Ground Snakes.)	Simotes Russellii	Bombay.
	Do.	Alibag.
	Oligodon subgriseus.........	
	Do.	
	Oligodon fasciatus	Bombay.
	Oligodon spilonotus	Do.
XII.—LYCODONTIDÆ (Harmless-fanged Snakes.)	Lycodon aulicus	Tanna.
	Do.	Bombay.
	Do.	Do.
	Do.	Do.
	Do.	Do.
	Do.	Do.
	Do.	Do.
XIII.—COLUBRIDÆ—		
I.—Group CORONELLINA (Ground Colubers.)	Cyclophis calamaria	Mahableshwar.
II.—Group COLUBRINA (Agile Colubers.)	Zamenis fasciolatus	Khandalla.
	Do.	Tanna.
	Cynophis malabaricus	Khandalla.
	Ptyas mucosus juv.	Bombay.
	Do. juv.	Do.
	Do. juv.	Do.
	Do. (head of adult)..	Do.
	Do. (head of adult)..	Do.

Family.	Genera and Species.	Locality.
XIII.—COLUBRIDÆ-ctd.		
III.—Group DRYADINA (Bush Colubers.)	None.	
IV.—Group NATRICINA (Amphibious Colubers.)	Tropidonotus quincunciatus	Bombay.
	Do.	Do.
	Do.	Alibag.
	Do. stolatus	Bombay.
	Do.	Do.
	Do.	Do.
	Do. Beddomii ...	Mahableshwar.
	Do. plumbicolor...	Khandalla.
XIV.—DENDROPHIDÆ (Tree Snakes.)	None.	
XV.—DRYIOPHIDÆ (Long-nosed Tree Snakes.)	Passerita mycterizans	Tanna.
	Do.	Bombay.
	Do.	Do.
XVI.—DIPSADIDÆ (Broad-headed Tree Snakes)	Dipsas gokool	Do.
	Do.	Saugor, C. P.
	Do.	
	Do.	
	Dipsas ceylonensis	Alibag.
XVII.—PSAMMOPHIDÆ (Desert Snakes.)	None.	
XVIII.—ELAPIDÆ (Venomous Colubrine Land Snakes)	Bungarus arcuatus	Bombay.
	Do.	Saugor, C. P.
	Do.	
	Naga tripudians	Bombay.
	Do. (head)............	Do.
	Do. juv	Born in Society's rooms.
	Do. (embryo, with tooth for cutting egg.)	
	Callophis trimaculatus......	Mahableshwar.
	Ophiophagus elaps (skin)..	Canara.
XIX.—HYDROPHIDÆ (Sea Snakes.)	Enhydrina bengalensis......	Bombay Harbour.
	Do.	Do.
	Do.	Do.
	Hydrophis diadema	Bombay Harbour.
	Do.	Do.
	Do.	Do.
	Do.	Do.
	Hydrophis robusta	Alibag.
	Pelamis bicolor............	Do.
	Do.	Do.
	Do.	Do.
	Do.	Do.
XX.—CROTALIDÆ (Crotali or Pit Vipers.)	Trimeresurus anamallensis .	Khandalla.
	Do. (head)	Do.
	Hypnale nepa	Ceylon.

Family.	Genera and Species.	Locality.
XXI.—VIPERIDÆ (Vipers.)	Echis carinata	Rutnagiri.
	Do.	Do.
	Do.	
	Daboia elegans	Bandora.
	Do. (head of 61 specimen.)	Hurda, C. P.
	Do.	Bombay.
	Do.	Do.
	Do.	Do.

NOTE.

It will be seen that in the Society's collection there are no specimens of the genera belonging to the following families :—

Fam. II.—TORTBICIDÆ (Short tailed Earth Snakes).
  ,, VII.—XENOPELTIDÆ (Iridescent Earth Snakes).
  ,,   X.—AMBLYCEPHALIDÆ (Blunt-headed Snakes).
    XIV.—DENDROPHIDÆ (Tree Snakes).
    XVII.—PSAMMOPHIDÆ (Desert Snakes).

Up-country members who are willing to assist the Collection, can have jars, containing spirits of wine, sent to them on application.

H. M. PHIPSON,
Honorary Secretary,
REPTILE SECTION.

PROCEEDINGS OF THE SOCIETY DURING THE QUARTER.

THE usual monthly meeting of this Society was held on Tuesday evening, 5th January, in the rooms at 6, Apollo Street. There was a large attendance of members. Dr. Macdonald having taken the chair, the minutes of the last meeting were read.

The following new members were then elected :—Lieutenant-Colonel Rowlandson, Captain Gerald Martin, Captain E. F. Marriott, Surgeon Horace Yeld, Miss E. Rich, Khansaheb Dinshahjee Dosabhai Khambatta, Rao Bahadoor Ragoonath Mahadev Kelkar, the Rev. Mr. Alexander, Messrs. W. M. Macaulay, A. C. Parmenides, Anthony Morrison, H. W. Jones and W. W. Squire.

The additions made to the Society's collections since last meeting were reported, as detailed below.

The Secretary reported that His Excellency Lord Reay had accepted the office of President of the Society. He also reported that he had been very successful at the auction of books mentioned at last meeting, having secured 13 separate works on Natural History, most of them rare and of great value.

Mr. Justice Birdwood proposed a vote of thanks to the Secretary, which was seconded by Mr. Sterndale, and carried *nem con.*

Mr. Sterndale then rose to propose a change in Rule VI., which runs thus :— "A president and two vice-presidents shall be elected from among the members resident in Bombay." He proposed that this rule should be amended so as to admit of the election of three or more vice-presidents, as in a place like Bombay, where many members are at certain seasons of the year absent, two are not sufficient. He also proposed that Mr. Justice Birdwood should be elected as a third vice-president.

On the suggestion of Dr. Bainbridge, these proposals were put separately, and, the first being seconded by Mr. F. N. Daver, was carried. Regarding the second, Mr. Kanga thought notice of it should have been given.

The Secretary said that notice of the intention to make a change in Rule VI. had been duly given, as required by the rules themselves ; but that he had not thought it necessary to give notice of Mr. Sterndale's intention to propose that Mr. Birdwood should be elected one of the vice-presidents.

Mr. Sterndale then rose to explain that his reason for wishing the matter carried through at this meeting was only this, that it seemed very desirable to have the governing body complete for insertion in the first number of the journal which he hoped would be in the hands of members by the 15th of this month.

Mr. Kanga at once agreed to this, and the motion, being seconded by Colonel Walcott, was carried unanimously.

The Secretary mentioned that the skulls presented to the Society by Mr. Shillingford of Purneah, which were acknowledged at last meeting, had since arrived and were now on exhibition in the room.

Mr. Sterndale proposed a vote of thanks to the Agent of the E. I. Railway for his courtesy in conveying the heads free of charge, which was seconded by Mr. Leslie Crawford and carried.

Mr. Justice Birdwood then exhibited some fruits of the *Ghela* (*Randia dumetorum*), a tree common at Matheran, which were inhabited by the larva of a butterfly, one of the Lycænidæ. The insect had in each case made a hole through the hard rind of the fruit and come out for the purpose of securing the fruit to the stalk with silk, lest it should fall. Some other curious phenomena were exhibited, and the meeting closed.

CONTRIBUTIONS.	CONTRIBUTORS.
Head of Jackal (with solitary horn between the ears)	Dr. K. R. Kirtikar.
Specimens of the Flora of Western Australia ...	Capt. O'Grady.
2 Walrus Tusks	Capt. W. Walker.
2 Australian Boomerangs	Do.
1 Live Koel (Eudynamis honorata)	Col. Bissett.
A quantity of fresh water fishes and crustaceans	W. Sinclair, C.S.
2 Bats	Do.
Live Octopus and Fish	Miss Walcott.

Minor Contributions—From Messrs. K. O. Campbell, Gibson, W. J. Essai, Rev. A. B. Watson, and Mr. L. P. Russell.

Exhibits—A live crested Hawk Eagle (*Limnœtus cristatellus*), by H. M. Phipson.

Additions to the Library.—*Malabar Fishes* (Day), presented by Mr. O. P. Cooper.

THE annual meeting of this Society was held on Monday, the 1st February, at 6, Apollo Street. Dr. Macdonald having taken the chair, the minutes of the last meeting were read and confirmed. The following new members were elected :—Dr. J. C. Lisboa, Miss Oliver, Miss R. Oliver, Colonel Goodfellow, Dr. H. Cooke, Messrs. W. Woodward, H. G. Palliser, J. Steiner. L. C. Balfour, B. B. Russell, John Chrystal, N. Spencer, P. Reynolds, C. Lowell, J. C. Francis, G. Oliver, N. H. Chowksey, and G. Manson.

The accounts for 1885 were put in. Mr. Sterndale proposed that Mr. F. G. Kingsley should be requested to audit them. The motion was seconded by Mr. Justice Birdwood and carried.

The Secretary proposed that a managing and financial committee should be appointed under Rule XIV., consisting of the following *ex-officio* members, with powers to add to their number :—The vice-presidents of the Society, the presidents and secretaries of the sections, and the secretary and treasurer of the Society. He also proposed that Mr. F. G. Kingsley should be appointed treasurer. The motion was seconded by Mr. Kanga and carried unanimously.

Mr. N. S. Symons proposed that the funds of the Society should be deposited in a bank and a banking account kept. This was seconded by Mr. Jefferson, and carried.

The additions to the collections and library since last meeting were acknowledged as detailed below.

Mr. Justice Birdwood proposed a special vote of thanks to Mr. G. W. Vidal and Mr. A. Newnham for their valuable contributions, which was seconded by Mr. Starling, and carried.

Mr. Sterndale then exhibited a curiously deformed horn of the Cashmere stag obtained by exchange from M. Dauvergne, on which he made some interesting remarks, showing how the deformity had probably been caused. He also exhibited and made some remarks on the skin of a tiger-cat. Mr. Aitken read a note by Mr. Newnham on the frequent occurrence of albinoism in Cutch, adding some remarks on instances from his own experience, tending to show that a sandy soil and dry climate exercised what might be called a bleaching effect on the colour, not only of birds and beasts, but of insects also.

Before the meeting closed the Secretary intimated that he had found a practical European taxidermist in want of employment, with whom he had entered into an engagement which he hoped would enable the Society to undertake any kind of work, such as curing skins, mounting heads and setting up birds, not only for members, but for other sportsmen and naturalists. All arrangements would, of course, be made through Mr. E. L. Barton, whose name would be a guarantee for the artistic finish of all work undertaken.

Contributions.—450 birds' skins, by Mr. G. W. Vidal, C.S.; skin of hamad-rayad (*Ophiophagus Elaps*), by Mr. G. W. Vidal, C S ; one snake (*Zamenis Fasciatus*), by Mr. G. W. Vidal, C.S.; 102 birds' skins, from Bhooj, by Mr. A. Newnham; large ant's-nest, by Mr. W. Shipp; one stuffed fish (*Barbus Carnaticus*), by Mr. H. M. Phipson, a quantity of small fresh-water fishes, by Mr. W. Sinclair, C.S.; three skins of *Capra Sibirica*, the Himalayan ibex, showing the colouring at three different seasons, by Mons. H. Dauvergne; one pigmy shrew, by Mr. H. Littledale; two hammer-headed sharks, by Dr. Hatch.

Minor contributions by Messrs. F. A. Little, John Chrystal, W. Shipp, W. Thacker, J. M. Mitchell, W. T. Smith, W. LeGeyt, K. M. Shroff, and D. E. Aitken.

Contributions to Library.—Birds of the Bombay Presidency (Barnes), by the author ; Encyclopedie d' Histoire Naturelle (Vol. 1-6), J. Poutz.

THE monthly meeting of the Society was held on Monday, March 1, in the Rooms at 6, Apollo Street, and was largely attended. Dr. D. Macdonald took the chair.

The following new members were elected :—Captain G. Wilson, Mr. D. Morris, Mr. J. H. C. Dunsterville, Mr. G. J. R. Rayment, Dr. Gaye, Mr. E. M. Walton, Major W. S. Bisset, R.E., Mr. G. H. R. Hart, Miss Hart, Mr. G. Fletcher, Mr. J. Anderson, Cap'. T. R. M. Macpherson, Dr. Henderson, Col. Westmacott, Miss Maneckjee Cursetjee, Mr. D. B. Maistry, and Mr. C. C. Mehta.

The following additions made to the Society's collections, since the last meeting, were duly acknowledged :—

126 species of ants and wasps, from Calcutta, by Mr. G. A. J. Rothney.

Several black bucks' heads and birds' skins, from Ahmedabad, by Colonel J. Hills, R.E.

Skull of hippopotamus, from Zanzibar, by Mr. F. D. Parker.

One snake (*Echis carinata*), by Mr. D. E. Aitken.

One Indian monitor (*Varanus dracœna*), by Dr. Kirtika.

One sarus crane (*Grus antigone*), by Mr. John Griffiths.

A quantity of mussels and sponges, Bombay harbour, by Miss Walke.

A quantity of polyps, Bombay harbour, by Mr. W. W. Squire.

Fresh water sponges, by Mr. W. Gleadow.

Four lizards, alive (*Urmastix hardwickii*), by Mr. R. M. Dixon.

Five snakes (*Silybura brevis, Chersydrus granulatus, Gongylophis conicus, Zamenis fasciolatus, Lycodon aulicus*), by Mr. H. M. Phipson.

Minor contributions from Messrs. H. W. Barrow, H. B. Mactaggart, J. Bristed, W. A. Collins, Thos Lidbetter, J. D'Aguiar, Major Kirkwood, and Captain Miller.

Additions to the Library.—Cyclopœdia of India, 3 vols. (Balfour), from W. Sinclair, C.S.; Asiatic Society Journal for 1885, from the Secretary, Calcutta.

Two panthers, two sambhurs, a cheetul, and a black buck, mounted by the Society's taxidermist for up-country correspondents, were also exhibited.

Mr. E. H. Aitken announced that, as he was about to leave Bombay, he was obliged to resign the position of Honorary Secretary, but expressed a hope that he would still be able to contribute to the Society's collections.

The Chairman proposed a special vote of thanks to Mr. Aitken for the energetic manner in which he had fulfilled the duties of Honorary Secretary since the establishment of the Society.

The vote, on being put to the meeting, was received with applause, and carried unanimously.

Mr. H. M. Phipson was then elected Honorary Secretary.

Mr. E. H. Aitken read an interesting paper on the classification of insects, pointing out the characteristics of the different orders, and describing their development.

The metamorphosis of the dragon-fly was most happily illustrated by the opportune appearance of one of these insects in the winged condition from the pupa state during the course of the lecture.

Mr. Sterndale exhibited some curiously formed horns of the Cashmere stag, showing a bifurcation of the bez tine, and a fine head of the musk-deer.

JOURNAL

OF THE

BOMBAY

Natural History Society.

| No. 3. | BOMBAY, JULY 1886. | Vol. I. |

A SIND LAKE,

By Capt. E. F. Becher, R.A., f.z.s.

Sind, as viewed on the map and as seen from the sea on approaching Karachi, has a most unpromising appearance; in the former case the Desert of Sind is written, and in the latter an apparently desert of deserts is seen, the few houses of Clifton, surrounded by sand hills, giving a greater aspect of desolation than if no signs of habitation were visible; but along the banks of the Indus which traverses the whole length of Sind are numerous jhils and lakes abounding in wild fowl.

The Manchar Lake, however, though communicating with the Indus, does not owe its existence entirely to that river; it is about 7 miles long and 4 broad; on one side are high barren hills of bare rock, and on the other an open cultivated plain stretching to the Indus, which is distant about 8 or 9 miles.

The lake itself is for the most part shallow and covered with water weed; the water is like crystal, and, looking down on the subaqueous forest through the clear shallow medium, brightened by the usual unclouded sun, it has always reminded me of a most perfect microscopical illumination of some opaque object, a beauty which a microscopist will understand. The surface of the lake teems with waterfowl. Mr. A. O. Hume says with respect to the coots : " I believe they would have to be counted not by thousands, but by tens of thousands. * * * In no part of the world have I ever seen such incredible multitudes of coots as are met with in Sind." This was written in 1873, but since that date Sind has been much opened out, and the Manchar Lake being easily accessible, the number of wild fowl has decreased. On three occasions I have spent about ten days on the lake. Living in a boat is much preferable to camping on the banks for any one to whom a bird is something more than a Hawk, Duck, or Snippet.

As an example of what sights gratify one's eyes in the early morning, it was no uncommon thing to see within a stone's throw of my boat the large and little cormorant, keenly engaged in catching their morning meal, at least two species of tern every now and then descending with a loud splash into the water, the common pied kingfisher hovering over the surface, stilts, one or two of the numerous graceful white herons or egrets, several black-tailed godwits, of course one or two of the numerous harriers which are perpetually sailing over the rushes, and two or three species of the smaller waders; other birds there were, but I think I have quoted enough; within a stone's throw is no exaggeration; no crouching behind a bush, or concealment was necessary on my part; they hardly paid any heed to my presence; on more than one occasion I have seen as many as three white-tailed eagles together almost within gunshot.

One of the methods of shooting wild fowl when required for the pot, and I am afraid often when not, is to be poled towards the numerous duck and shoot at them sitting on the surface of the water at long ranges; it is remarkable how they appear to know the exact range of an ordinary gun, but a choke-bore at present they do not understand; their almost invariable practice is to let you approach within 70 and 80 yards before they take flight.

On the banks are some fishing villages; great numbers of fish are caught by driving them into a net; this operation is accompanied by the most deafening and prolonged noise; if fish can hear, they would hear this; on the front of each boat is a rocking wooden tray in which is a copper dekshi; this tray is perpetually worked, varied with beating the deck with a short stick, the boat itself being rocked; a band conductor, as I will call him, as he seems to regulate the noise and movements, stations himself in a boat at the mouth of the net; it is no uncommon thing for these fishing boats to have a long perch, on which are seated various species of herons and egrets, and cormorants, or else, perhaps, a pedlican is standing on their boats. Mr. Murray says that they use these birds as decoys and sew up their eyes; in the case of those I have examined I am glad to say I have never seen this latter cruelty perpetrated.

The natives are adepts at spearing fish, which, when the fish are at some little depth is no easy matter; on account of the refraction, part of the equipment of every boat is two or more spears, and a stone on which to sharpen the points.

I always used to look forward to evening flighting, not only from a sporting point of view, but on account of the bird life which is always to be seen on these occasions; this shooting was always done from a boat concealed more or less amongst the reeds. I will take from my notes an account of an evening's flighting at the end of February last year. " About 4-30, I took up my position amongst the high reeds. The first to come over are one or two stragglers (duck), and then the usual enormous flocks of duck pass by, flying high over head from the direction of the Indus, the first intimation of their approach being the rushing noise caused by their wings; after this, or perhaps a little before, some large flocks of glossy ibis flying slowly in a single undulating line pass close by; one slowly unfolds one of its long legs and leisurely scratches its head, the whcle operation appearing very ludicrous; all the time one or two harriers hunt leisurely over the reeds ready to pick up any wounded victim to my gun; a gull or two pass over, especially noticeable is the large black-headed *Larus ichthyætus*, then comes a flock of graceful small white egrets; on one occasion I shot one for identification, which turned out to be *Herodias intermedia*; I also watch with interest the fishing of the blue kingfisher *A. ispida*, and perhaps *A. bengalensis*, and the pied kingfisher *Ceryle rudis*. (I might also have seen the lovely *Halcyon smyrnensis*, but as I am transcribing from my notes on this particular occasion, I did not.) Many wagtails of two or three species flit about the reed-covered surface of the water; the hoarse loud note of the Reed warblers, *Acrocephalus stentorius*, is constantly heard, but although close to me, I can only occasionally catch a glimpse of one amongst the reeds; the little warblers (*Phylloscopus tristis*) flit rapidly in and out amongst the rushes, and if I do not move, they allow me to admire their ceaseless activity almost within an arm's length; as the evening gets on, the croaking of the frogs and chirping of the grasshoppers (?) keep up a perpetual monotonous concert with the splashing and cackling of the noisy purple gallinules; cormorants, both great and small, fly past; (in the case of one I shot, the small cormorant was *Graculus javanica*, but in Mr. Murray's *Vertebrates of Sind* I see that both *Graculus sinensis* and *G. javanica* are common Sind species, the former being distinguished from the latter by having *no white thigh or cheek patch*; I did not know of this distinction at the time, so was not on the alert to discriminate between the two species); then I see a few curlews, a flock of crows, and flying close to the surface of the water a flock of *Hirundiniuæ*; they are gone too quick for identification, but doubtless *Cotyle sinensis* ; and then come the duck, but I do not see the cloud of them which last December used to rise from the lake as it were simultaneously, passing overhead in varying numbers ; in a quarter of an hour or so the flight is over, darkness has set in, and all is still save the croaking frogs and the chirping insects."

I have mentioned above that *Alcedo ispida* and perhaps *A. bengalensis* are to be seen ; but I must confess that I am fairly puzzled with *Alcedo ispida, A. bengalensis,* and a small form which Mr. Hume says : " ＊ ＊ compels me to identify it with *ispida* rather than *bengalensis*."—(See Stray Feathers, Vol. I., p. 168.) In no book that I have seen is the difference between *A. bengalensis* and *A. ispida* clearly pointed out. I have four skins of Sind blue kingfishers before me as I write: three seem to me almost the same, except one which is not so long and whose bill is a trifle stouter than the other two ; these I refer to *ispida*, but the fourth is much smaller and much brighter; its length is 5·75, bill at top 1·44, bill from gape 1·87, wing 2·65 ; the bill is blackish brown except at the base of the lower mandible, which is beneath reddish ; the ground colour of the head is very dark brown ; the throat is white and the rest of the under parts ferruginous, but on the breast the ferruginous feathers are tipped with faint light blue ; it is male, and was shot at the Manchar Lake on the 15th December 1885.

As regards the geese and duck, on the last occasion I visited the lake (Dec. 9, 1885) geese, duck and other wild fowl were conspicuous by their absence, and I believe throughout Sind ; on this occasion I only saw a few grey lag geese (*A. cinereus*), but in February of the same year I have no note of this species, but the barred-head goose (*A. indicus*) was extremely abundant.

The Large Whistling Teal (*Dendrocygna fulva*).—I shot a few in December, but none in February ; they are very slow flyers, and when one of their number is shot, they often circle round it, constantly uttering their whistling cry; their feet and tarsus are proportionally very large, and altogether they give any one, who remarks individuality in other than the human species, the idea that they are half-witted.

The Ruddy Shelldrake (*Casarca rutila*), more generally known as the Brahminy, is common ; its hoarse croak is often heard as it flies overhead ; I cannot agree with the statement in Mr. Murray's Vertebrates of Sind that " they are extremely shy and wary birds," and, as

Mr. Reid in Game Birds remarks: "It will not only keep a sharp look-out on its own account, but will fly along the jhil side before the gunner, uttering its warning note and put every bird, on the *qui vive*." I have always found it a slow clumsy bird, easy to approach. I was very amused on one occasion watching a Pariah dog trying to approach one in some deep mud; the dog with an unconcerned manner, as if Brahminy duck was the one thing in this world which it had the least thought of, the duck as if a dog trying to catch it was an equally distant thought; the dog at last manœuvred till it was quite close and was evidently heedless of the proverb "First catch your hare before you cook it;" but then the Brahminy flapped away a few paces; then the same manœuvres were repeated to the evident amusement of the bird and the annoyance of the dog; how long the dog would have pursued in this wild goose or more correctly wild duck chase I cannot tell, as I was tired before the dog was; walking on put a stop to any more manœuvres; this duck and the former are considered not fit for human food; a brother officer tried a young Brahminy on one occasion and ate some of it with relish; he also had a whistling teal cooked; which he and another friend pronounced good; I have never eaten the former, but I have attempted to eat a little of the latter; I shall never do so again.

The Shoveller (*Spatula clypeata*) is very numerous; as a bird for the table it also has a bad reputation, which, no doubt, is frequently well deserved, as it is a foul feeder and delights in any dirty pool; but those I tried at the Manchar Lake were not bad eating.

The Mallard (*Anas boschas*).—Last December I think this was almost the most numerous species on the lake; in February I only shot two in about seven days' shooting.

The Gadwall (*Chaulelasmus streperus*) is also very common.

The Marbled Teal (*Chaulelasmus angustirostris*) very common. When flying, on account of its proportionately large expanse of wings, it appears a much larger bird than it is.

The Pintail (*Dafila acuta*), another very common species.

The Widgeon (*Mareca penelope*), not very common; I only shot one last December.

Both the Common and Garganey Teal (*Querquedula crecca* and *Q. circia*) are common, especially the latter; none of the males which I shot of the last species during my last December visit had made any attempts to assume the male plumage.

The Red-crested Pochard (*Fuligula rufina*) and the Tufted Duck (*F. cristata*) are fairly common, especially the latter. I did not shoot a single one of either of these ducks last December, nor did I observe any, nor did I see any pochard (*Fuligula ferina*) at that time ; I have only a note of it forming part of my bag last February, but whether common or not is not mentioned.

The White-eyed Duck (*Fuligula ferina*) is common.

At the latter end of the season, when the water has fallen, Snipe common and Jack are numerous in favourable places round the edges of the lake.

On the babul-fringed banks of the canal from Sehman I secured a male and a female of *Passer pyrrhonotus* ; this is an interesting bird from having been rediscovered by Mr. Doig in 1880, not having been recorded in India for forty years previously—See Stray Feathers, Vol. IX;

As regards the other animals inhabiting the lake, which particularly attract notice, amongst the fish there is a fresh-water pipe fish in considerable numbers ; in fact, it is almost impossible to look down into the water without seeing several of these gliding in and out amongst the weeds ; the natives never seem to catch it ; there is also a fresh-water prawn which to the eye uneducated in entomostracan lore appears similar to the well-known marine form. Mr. Murray informs me that it has not as yet been properly identified.

There are several species of fresh-water shells, one,—a fresh-water mussel,—is very numerous ; there is another form of large bivalve, which is unknown to me ; Limnœa sp. (?) is also very common with a pink variety ; Sphœrium sp. (?) fairly numerous ; a smallish Planorbis sp. (?) is met with on the weeds, but not in any great numbers ; Paludina sp. (?) is very common.

As regards the vegetable kingdom, one of the commonest sights is to see a number of naked women digging up from the mud the roots of the lotus, whose broad leaves cover the water in places, and afford a convenient standing ground for snipe, as I found to my cost, when working the neighbouring snipe ground ; these roots seem to be rather highly prized as a vegetable; I tasted them, and they had the flavour of parsnip, but were rather stringy, as they seem chiefly made up of a number of fine silk-like fibres. But for the present I have said more than enough ; if I were to write of all I saw at the lake, I am afraid the journal of the B. N. H. S. would scarcely contain it.

E. F. BECHER, Capt., R.A., F.Z.S.

NOTES ON THE WATERS OF WESTERN INDIA.

Part I.—" British Deccan and Khandesh."

By a Member of the Society.

The following rough notes on the waters of Western India are written " gryphonibus puerisque," and I do not suppose them to contain much original matter of any scientific value. It is hardly necessary to say that I have drawn freely upon the standard works of Drs. Jerdon, Nicholson, Day, and Gunther, but more special acknowledgment is due to later and less known local writers, Mr. Wendon, C.E., Dr. Fairbank, Captain Butler, and other officers who contributed to the *Bombay Gazetteer* and the Reports attached to the Bombay contributions to the Fisheries Exhibition. Even of my own observations, the memoranda used in these notes have mostly been put at the service of the officers who compiled these last-named publications, or used in a lecture delivered before the Royal Asiatic Society. For the Indian angler, Mr. Thomas's " Rod in India" stands by itself ; and whoever wants to catch fish in this country ought to read it, and not depend on my incomplete remarks.

As but few Europeans on this side of India are much in the way of sea fishing, I shall begin by describing the fresh waters of the Presidency, which are divided between four very well-marked regions.

The first of these is that of the Deccan and Khandesh. All along the Western Ghâts a number of torrents rising very close to their scarped edge flow eastwards ; generally, at first, with a good deal of southing. Within a very few miles of their sources these unite to form rivers, the beds of which a good deal resemble those of salmon rivers in Northern Europe ; but their streams differ from these in an important particular. Instead of the alternate rise and fall which make European angling a speculative pursuit, we have here three or four months of continual flood, while for the rest of the year each river becomes a chain of pools connected (if at all) by a very insignificant current. Another matter very important to the fish is that this region of torrents and moderate-sized rivers is also one of rice cultivation carried on in small pond-like fields called *kasars*, through which a great deal of the water from the hill sides must pass before it reaches any definite channel. Below the rice region these rivers generally flow through wide valleys for from 50 to 100 miles before reaching the great plain of the Deccan. Their course (as will have been understood from my comparison of them to salmon rivers) is much diversified with rapids, sometimes even with considerable falls, with gravelly shallows, and with long pools and reaches. These latter occasionally have alluvial banks and muddy bottoms, but more commonly the bank is rocky; the bed of the same nature, with a good deal of gravel; and the water clear throughout the fine weather, that is, from October to May inclusive.

There is hardly a single river of importance that is not crossed by at least one ancient or modern irrigation weir; and on some there are many weirs, all of masonry, sometimes very lofty, and in no case that I know of provided with any sort of a fish-ladder. As many of the tributary torrents as have any stream during the whole or part of the dry season are crossed by many little dams, usually built for the season only, of wattles, mats, and mud or gravel, but sometimes they also are permanent dams of good stonework.

As each group of these rivers debouches from its gradually widening valleys into the great plain of the Deccan, some one of them, like Aaron's Rod, swallows up the others; and from this point to the eastern boundary of the Presidency its course is generally a huge trough about 100 feet deep and half a mile wide, bottomed alternately with sand and mud, and rarely crossed by a bar of basalt, over which the river falls in rapids or a cataract.

Except at such places the banks are usually of stiff alluvial soil, scarped on the outside of each curve of the stream, where it runs deepest and strongest, but sloping gradually on the inside of the curve to wide sandbanks bordering on the "dead water."

The streams which unite to form the Bhima, most of which rise in the Poona District, illustrate the above description well enough; but the finest falls on any *large* river easily accessible from Bombay are those on the Godavery at Phultamba.

Before dismissing the Deccan rivers it should be added that each of them after leaving this Presidency is barred by great irrigation works, which completely prevent the ascent of fish from the sea from their lower waters.

Besides its rivers, the Deccan has a considerable number of artificial lakes and ponds, or, as we call them, tanks. Some of these, especially those at Khadakwasla, near Poona, and Ekruk, near Sholapur, are of considerable size, and a good many, even of the lesser, are perennial. But the greater number are reduced to mere

puddles, or entirely dried up annually, even in ordinary seasons·
Of natural lakes there is not one.

Khandesh, for the purpose of these notes, may be classed with
the Deccan, which it resembles in its geology and hydrography ;
and though its great river, the Tapti, flows into the Arabian Sea,
instead of the Bay of Bengal, it has only one tributary of importance
(the Púrná) that does not rise in the Western Ghâts, or in their
great spur, the Satmalla Range. · Rivers and tanks in these two
neighbouring regions resemble each other, even as Fluellen's waters of
Macedon and Monmouth. It is true that instead of "salmons in
both," " there is salmons in neither ;"* and it is now perhaps time
to consider what there is instead of salmons.

Nearly all the fishes of any importance belong to two families,
namely, the Cyprinidæ, or Carps ; and the Siluridæ, or Catfishes.

Probably no writer on Indian fishes, except a professed ichthy-
ologist, can escape from beginning with " the Mahseer." As a
matter of fact, although it would not be correct to say that
there is no such fish as a mahseer, there is certainly no
fish that has an exclusive right to the title, and it is not a genuine
native name for any fish in our present province. A certain group
of Indian barbels differ from the English representatives of that
genus in preferring troubled waters and a highly predatory existence.
They will eat, indeed, whatever they can come at, from a fly to a
wild fig ; but what they like best, perhaps, is a little fish, no matter
of what sort, even if their own. This frame of mind and palate fits
them particularly for the purpose of the sportsman, and wherever you
find him in India, he and his native assistants will be found calling
some of these predatory barbels " Mahseer" or " Big-head." Even
where the term is vernacular, viz., in Hindustan, it varies in local
application, and still more in the Peninsula.

Naturalists, however, have generally agreed in appropriating the
title to the giant of the tribe, " *Barbus tor*," of whom all that I can
say here, unfortunately, is that within our present area he is not at
all a common fish; and when found, not often a very large
one. The reason is not far to seek. The great rivers of the Himalayas, in which the true " Mahseer" thrives, are fed by rain
and melting snow at different seasons to an extent that makes them
and their upper tributaries perennial. Many of those of the extreme

---

* NOTE.—The " Rajputana trout" (*Barilius bola*) and the " Himalayan trout"
(*Oreinus*, several species) are not found in this Presidency. Both are Cyprinidæ.

south of India, where also this fish flourishes, get the benefit of two
monsoons; and in both cases the upper streamlets run from lofty
mountains through, at first, uninhabited jungles of great extent,
where spawning fish and descending fry are pretty secure from
their worst enemy—man.

The streams of the Deccan, on the other hand, are full for only
three or four months, and even at that season the sources of almost
every one of them, as far as the barbels are concerned, are, and
have been for many generations, in rice-fields, out of which few
spawning fish, and not many of their fry, escape alive. All the
circumstances are against large fish like *Barbus tor*, with a taste for
high spawning grounds, and in favour of species more moderate in
size and aspiration, though otherwise of very similar appearance and
habits. These are generally known to the natives as "*Kawli Masa*"
or "scaly-fish" from their large scales. If I remember right, the
allied Burbot has a similar local name on the Rhine. Dr. Fairbank
gives "*Mhasala*" or "Buffalo-fish" as a Mahratta name for *Barbus
tor*, and mentions one as 3⅓ feet long, one foot *high* (!), and weighing
42 lbs., much the largest I ever heard of in these waters. As
regards the value of the whole group for the table, all I can say is
that I never tasted a Mahseer of any one else's killing that was worth
putting a fork to. What I kill myself are (of course) good fish all
round. They will all sometimes rise at a fly or a spinning bait (dead
or artificial), but live bait is certainly the most killing. The name of
"Indian salmon" is an absurd misnomer for these or any other
Indian fishes; a Mahseer no more resembles a salmon than a Buc-
caneer might an English naval officer.

Next after the Mahseers come the Labeos, or *Rahu* or *Roho* fish,
named by Hindu fancy after the mythical dragon who causes eclipses
by swallowing the sun. The type of the *genus*, perhaps, is Labeo
Rohita, the "*Roho* fish" proper, called in Mahratta "*tambada
masa*" or "copper fish." The name "*Roho*" is as much knocked
about as that of Mahseer. These Labeos are easily distinguished at
the first glance from the Indian barbels by their longer form and very
peculiar mouth, set under the snout, and furnished with thick warty
lips, convenient for grazing from above on water weeds, which, with
perhaps some insects and snails, form "the chief of their diet."
They like still and muddy water; in this resembling the European
carp; and I should certainly have called them "Indian carp" in
this paper if Mr. Thomas had not most unfortunately appropriated

the title to an omnivorous fighting barbel closely allied to the Mahseer and actually called Mahseer by Europeans in our province. *Factum valet quod fieri non debuit*, the Rohos must go without an English name. In net-fishing throughout our province they are usually the largest fish in the net, but are very apt to escape by jumping over it in fine style. I have more than once seen one knock a man down and go off over his prostrate body, and have got good sport by wading behind the net with a spear and striking them in the air. The best baits for them are paste, earth-nuts and gram. Worms are so scarce in this country that one can hardly count them among available bait, but when you *can* get them, hardly any Indian fish will refuse them. If any gentleman despises bottom fishing, let him try for a Roho with fine tackle (coarse tackle is of no use), and if he hooks one, he will find the play much more like that of a salmon than a Mahseer's; and the fish, moreover, very much better for the table. With a little trouble they can be kept alive for a good while, and even when dead do not quickly become stale.*

After the Mahseers and Rohos there are no Cyprinidœ of any account either for sport or for the table, though several small sorts, such as Chela, Rasbora, and Barilius, can be taken with a midge-fly or small bait and trout rod, and fried in rows upon a bamboo splinter, after the fashion known to mofussil house-keepers as "Havildars and twelves." If small enough, they can then be eaten, bones and all, and are no bad variety in the monotonous bill of fare of a camp.

The next family, the Siluridœ or catfishes, though not so numerous in individuals, are quite as often "in evidence," as several of them are much better eating than any Indian Cyprinoid. They are all scaleless, and most of them have a "dead fin" behind the great back fin like a salmon or trout. The commonest and best for the table is the "*Padi*," or "*Shiwara masa*" (*Wallago attu*), the Boalli of Upper India. Dr. Fairbank gives "*Padi*" as a name for *Silundia Sykesi*, another catfish, much handsomer, and possessing a dead fin, for which Sykes himself gives "*Pari*" and "*Sillun.*" *Wallago attu* grows to a great size, bites well, and shows good fight. On one occasion I had played one almost within reach of the landing net, when a second of about equal size rushed

* NOTE.—Shah Jahan or his father, I forget which, gave a horse and a village to a lucky angler who brought him a fine "Rahu machi." The story is in Elliot: *auctore Imperatore ipso.*

up, laid hold of the captive, and carried him off into deep water, where, after a few minutes, the fine tackle gave way.

The terms *Singhala*, *Singhata*, &c., signifying " Horn-fish," are applied by Mahrattas to several catfish with long feelers, mostly of the genus Macrones. These generally give fair sport, and are good eating. The best way of angling for any of them is to use a live bait in the evening, when they leave the deep water, and maraud along the banks, or near the surface. Failing such bait, fresh raw meat answers fairly well. It is good to shoot some wild bird or kill a chicken beside the river bank, and bait with warm flesh, as all carnivorous fish are strongly attracted by the smell of blood.

In handling the catfishes it is necessary to be very careful, as several species are provided with formidable spines, to say nothing of numerous and sharp teeth; and the wound of either is apt to be very painful, and takes long to heal.

The larger species are sometimes known to sportsmen as " Fresh-water sharks " from their size, temper, and well-furnished jaws.

After these there is only one family of sporting fish left to name, viz , the walking fishes or *Ophiocephalidæ* (snake-heads), commonly called " *Murrell*." These are long fish, something of the shape of a ling, whose head is fancifully supposed to resemble that of a snake, whence the scientific name.

The Murrells are known to natives in the Deccan by that name, but elsewhere in this Presidency as *Dhak*, *Dhakru*, or *Dhok*. They are chiefly remarkable as air-breathing fish, a quality which enables them to live for many hours out of water, and even to move for some distance over land, wriggling and crawling with their flapper-like fins, whence their English name. They cannot, indeed, live altogether under water, but must rise to the surface occasionally to take in fresh air ; and they like to lie at the top with their nostrils exposed and breathe air for long periods together. To do so in the centre of a stream or tank would expose them to many enemies ; and the Murrells accordingly lurk in thick beds of weeds, or under overhanging roots or rocks on the bank, where they lie half erect in the water, breathing air and looking out for wind-falls. They are said to have subaqueous burrows, but these, in the nature of things, they cannot use for any long time to-gether, and in my opinion they pass most of their lives at the surface, but so skilfully concealed that they are seldom observed.

In such a position they can sometimes be caught by dropping a frog, grasshopper, or the like, upon the water close to them ; but this is usually very difficult to do without being seen by the fish. At night they leave their lurking places and cruise for prey near the surface, and then they are often caught with trimmers baited with live fish or frogs, or in favourable places with the rod, using for bait the smallest possible fish, frog, tad-pole, or even fresh raw meat. I once caught over two dozen of a small species with the rod in one evening with the latter bait. The Murrells are said to be monogamous, and, in fact, patterns of domestic virtue until their young come of age, when the parents turn them out to seek their fortune ; and *eat the laggards*. All of them are good eating when in season, but at other times muddy flavoured. The same is the case with the catfishes, and this is usually accounted for by the difference of waters. My own experience is, however, that these fishes, like salmon, are often good eating even when taken from still and muddy waters, and earthy flavoured in the clearest streams. I have no doubt that it is with them, as with the salmon, a question of season.

In some rivers considerable numbers of Murrells are shot, as they rise to the surface, with bullets or with barbed arrows. The arrow-heads are loosely set, but connected with the shaft by a line wound round it. The archer plunges into the water, recovers the floating arrow-shaft, and hauls in the fish by the line. The mere shock of the bullet on the water will often stun a fish without actual contact.

The last thing to be said about these interesting fish is that they have the power of lying asleep in the mud of dried-up tanks until the return of the rains,—a power shared by several other fish of this region, especially by a queer-looking creature, called " *Wambh,*" " *chalát,*" and " *chambáre*" (" tanner-fish "), *Notopterus kapirat.*

. True eels (*Ahir*) are not very often caught in the Deccan, partly because they are really not common, but still more because the fishing gear of that country is unsuited for their capture. I only once saw one caught, *viz.*, at Phultamba, on the Godavery, a famous neighbourhood for fish. My Portuguese cook refused to cook it on the ground that it was " all same like ishnake." There is only one species, *Anguilla bengalensis,* which grows to at least 5 lbs. weight.

No prejudice attaches, however, to the spiny eels, called commonly "*Bhám*" and "*Wambhat*," strange-looking fishes with rows of prickles and long "trunk-like" snouts. They are very good eating, but of no importance from a sporting point of view, though I have seen my servants catch them on hooks baited with raw meat.

Besides the lesser Cyprinidæ mentioned above, several fresh-water herrings will take a trout-fly, giving a good deal of amusement in a small way, and these are all good for the table in the form of "Havildars and twelves." Along with these is sometimes caught the queer-looking fresh-water garfish (*Belone cancila*), called in Mahratti "*kutra*" or "dog-fish," probably from its greediness, or from its long well-armed jaws. It is exactly like the garfish of European seas, living mostly close to the surface, and very fond of skipping over any floating stick or straw. In our present province both game and meat are often very scarce, and after many days' diet of tough mutton and tougher "moorghies" in a bad climate, a very moderate dish of eatable fish is a welcome luxury.

Setting aside nets and traps, it may be said that the main points for the angler to remember in such waters as I have been describing are to use a trout-rod for small fish, a salmon-rod for the large ones, the finest line he dares, and the smallest hooks on the strongest gut that he can get. Even in spinning he should never use treble hooks, because almost all the fish he looks out for, except some catfishes, have small mouths; and the mahseers, though they have no teeth in their mouths at all, have such power of jaw that they can break anything that offers resistance, as a treble hook does. If further information is required, the best of it is to be got in Lieutenant Beavan's "Freshwater Fishes of India" and Mr. Thomas's "Rod in India."

I repent that I have omitted to notice one handsome genus of carps, the *Cirrhinas*, which are very good eating, and would probably, if one could get them to take either a fly or bait, give better sport than any other Indian fish, as they have certainly no equals in grace of form and motion.

Although the fishes have claimed precedence in remarks upon their own element, their possession of it is disputed by many other creatures. In our present province, excluding man, only one of these is a mammal, viz., the Otter (*Lutra nair*), called in Mahratti "*Ud*," "*Lad*," and "*Pán-Manjar*," (*i.e.*, "Watercat"). I once heard a Kashmiri Pandit call one "*Ludra*," which comes close as can be expected to the Latin and Greek. This animal is far more common in the neighbourhood of the ghâts than is supposed by most sportsmen ; but being very shy, and of nocturnal habits, is rarely seen. If, however, one follows up any river near Poona, for instance, in the early morning, one is pretty sure to come on his unmistakeable " seal " on a mud bank, and very likely on the remains of his supper. The otter of the Deccan is much smaller than in Upper India and Sind, though classed as the same species.

Aquatic birds are more numerous. I have never seen any of the fishing eagles in the Deccan,* but the Osprey is not very uncommon, and the chestnut and white " Brahminy Kite " does a little fishing. He cannot go under water like the Osprey, but picks up small fish from the surface. The fishing owls (Ketupa) are very rare here, being essentially forest birds. Specimens of two species were sent from this Presidency to the Fisheries Exhibition, but it is not stated whence they came. Of Kingfishers, 5 species are found, as follows :—

(1) The Large Blue Kingfisher, *H. Leucocephalus* ;
(2) The Lesser Blue Kingfisher, *H. Smyrnensis* ;
(3) The Least Blue Kingfisher, *Alcedo bengalensis* ; and the
(4) Pied Kingfisher, *Ceryle rudis.*

> The two last are the commonest, especially in the open plains ; the others prefer wooded streams, and vary their fish diet a good deal with grasshoppers and the like. *Halcyon smyrnensis,* indeed, seems almost independent of water, wherever there is woodland. The Pied Kingfisher is the most conspicuous and best known from its habit of hovering over open water and dropping like a stone upon its quarry. I heard on good authority of its attacking in this manner a dog that had passed too near its nest in a bank.

(5) Colonel Sykes records the rare and beautiful Three-toed Purple Kingfisher (*Ceyx tridactyla*) from this region. The whole tribe are known to Mahrattas as " *Dis*" and " *Kilkila.*" They generally build in holes ; but once in Sind I found *Alcedo bengalensis* breeding in a very rude pendulous nest in the grassy over-hanging bank of a canal. The young were destroyed by a flood. I fancy that this kingfisher was not the original architect of the nest.

* The white-tailed sea-eagle (*Poliœtus ichthyœtus*) is recorded from Dharwar.

The common and Demoiselle cranes do not touch fish or spawn, and the large Sáras crane, which is accused of doing so, is very rare in the Deccan and Khandesh. It is not likely that any Plover can interfere much with fish or spawn, though I once saw a common "Did ye do it" (*Lobivanellus goensis*) catch and eat a small fish. It is, indeed, the only Plover which haunts the waters of our present province in important numbers. *Esacus recurvirostris*, the great Stoneplover, is found here and there in the beds of large rivers, and perhaps may eat spawn, or even fry occasionally, but its main dependence is on insects and crustacea, with a few shellfish.

Of the *Longirostres*, the snipes and their allies we have, though in no great numbers ; the "full" snipe, "painted Jack," and "pin-tail" snipe ; the greenshank, several sandpipers, and stints ; curlews and whimbrels (both rare) and the stilt (*Himantopus candidus*). This bird and its tribe would probably devour fish and spawn, but I do not know of any positive evidence against them ; and most of them can plead *alibi* here, being cold-weather visitors only. The stilt and greenshank, though not very sporting birds, are very good for the table.

The coots, waterhens and rails are chiefly represented here by the bald coot, the European waterhen, and the white-breasted waterhen, *Gallinula phœnicura*. The second of these is much accused in England of eating fish spawn ; the first nowhere, I think, and the last seldom enters the water of its own accord, though usually living near it. It is, in fact, a bird rather of the bank than of the river, and I have shot one 20 miles from any bigger water than a well. All three breed within this region.

The next tribe, however, the *Cultirostres* : Storks, Ibises and Herons are mostly very much dependent on the water. Their chief, the Adjutant, can, indeed, do well enough without it. He is rare in the Deccan, much less so in Khandesh ; but he fishes rarely or not at all. The fine black-necked stork (*Mycteria australis*) is rare, and so are the black and the white stork (*Ciconia nigra* and *alba*), both of which are northern birds that hardly get so far south as the Deccan, even in the cold weather. Even the name of the former is here appropriated by the resident white-necked stork (*Ciconia leucocephala*), which breeds here in trees in the rains, and is very common, foraging both on land and water, but chiefly on the edge of the latter. It eats plenty of fish, still more frogs, crabs, and tadpoles, lizards, grasshoppers, and, it is said, sometimes snakes, and even field mice.

This fowl of a mixed diet is sometimes eaten himself by the lord of creation, under the name of "beefsteak bird" for a change. So is his frequent neighbour, the Pelican ibis, (*Tantalus leucocephalus*), who lives in much the same way and in the same places, and is not uncommon here. The white ibis is found on the larger rivers, often along with its relative, the spoonbill; neither is common, and neither can eat many fish, though they probably do not spare spawn when they find it. Both are eatable, though coarse in flavour. The shell ibis is almost unknown; the glossy brown ibis rare; and the red-headed black ibis has hardly the habits of a water bird at all. I regret to say that upon slight temptation he becomes a mere scavenger; but in places where he cannot get at dirt, he is, though coarse, quite eatable.

These ibises have intruded themselves wrongfully between the storks and the herons, which are numerically exceedingly abundant. Up to the present we have had to deal with no creature, except the osprey and kingfishers, which can be called a mere enemy of the fish. For the otters and the piscivorous birds mentioned above (with the exceptions given) destroy more frogs, water insects and crustacea than they do fish, and all these are deadly enemies of fish spawn and young fry.

The herons, however, and most of the birds remaining for notice, subsist almost entirely on fish.

The common grey European heron is found on all the rivers and tanks, and requires no special notice. The great Malayan herons, *A. Goliath* and *A. Sumatrana*, are not, I think, found in this Presidency, though Sir A. Burnes figured something like *A. Sumatrana* from Sind. A bird somewhat allied to it, the purple or grass heron, is found on a few weedy tanks in the Deccan, but is not common; nor is the queer-looking night heron, which, though its nocturnal habits keep it a good deal out of sight, generally lets one know of its whereabouts by its peculiar and often repeated cry.

The egrets are numerous, and first amongst them is the great egret (*Herodias alba*), valuable for the long feathers of its back. These are at their best in the early breeding season.—May, June and July. Their growth coincides with the change of the beak from yellow to black; and the plume-hunter should therefore not waste his shot on an egret with a *yellow* bill. The same is the case with the lesser

white egret, whose plumes, though, of course, smaller, are still
worth having.

· The cattle egret, with his buff plumes, can hardly be counted a
water-bird, and the bittern is rare; but the little paddy bird is
really one of the "features of the landscape" all over India. You
find him on every stream and pond picking up fish, tadpoles,
crabs and what not, and occasionally swimming, or rather floating.
He does not, as far as I am aware, ever *fish* beyond his depth. The
sudden change of this little heron from a grey bird to a white
as he flies off is a real transformation; and his moult from grey
to purple and white is quite a hard thing to get young naturalists
to believe in. The bittern is rare in our present province; and
it would take up too much time to go further into the history of
the smaller herons, with which, indeed, this is not a favorite region.

Of the great tribe of ducks and geese there are hardly any that
will not eat fish spawn whenever they can get it, and few that
do not occasionally pick up small fish, but the latter are not
the principal food of any found here, and during the rains,
which are the great spawning season of the fish, you might go all
through the Deccan and Khandesh without seeing a single duck
or teal of any description, unless on some remote tanks which are
favoured by the *nukta*, or black and white goose, with its queer
bottle-nose, its duodecimo-edition, the cotton teal, and the bay-
coloured lesser whistling teal. Dr. Fairbank and myself have observed
the larger whistling teal in the Ahmednagar District, but I think
it is only a cold-weather visitor there, and it is certainly very rare.
It does, like the three above-mentioned, breed in other parts of
India. The whole four are very poor eating in the cold weather,
when the migrant ducks are most numerous and in best condition;
but they improve much in flavour in April and May, just when the
northern visitors are not to be had. This is easy enough to
understand if we consider that the northern waterfowl begin to
breed in late spring or early summer, and have got through the
trouble of raising their families in July and August. From that
time till the next spring they think of nothing but filling their
stomachs, and though they fall off a little in condition during
their long flight across the mountain barriers of India, they soon
recover it. The few snipe, for instance, that remain here till
April, which are celibate fowls with digestions unimpaired by any
affection of the heart, get to be mere balls of fat, and a tailor might

knock them down with his goose. Contrariwise, the late snipe in the British Isles, birds with such strong family affections that they marry on the spot instead of going to Norway and Russia to do it, are almost unwholesome.

To return to our Indian ducks. These mostly breed from July or August, and at Christmas they have hardly yet recovered from their domestic exertions. But by April and May they have fully regained condition, and the young birds have acquired their full size, or nearly. The first in rank of the migrant ducks is that very eccentric bird, the flamingo. It is likely enough that some readers may be surprised at my calling it a duck at all. However, if any gentleman in that frame of mind will shoot a flamingo, and then compare its feet and the inside of its bill with those of the nearest duck, he will probably begin to admit that there is some reason for doing so. If the experiment is followed up by keeping it fifty or sixty hours in its feathers, plucking it, and roasting it, he will probably become a convert. Skinned birds, and especially birds kept after skinning, taste very different from those simply plucked. A skinned teal, for instance, is quite unrecognizable.

Our cooks have an execrable habit of plucking birds many hours before they cook them, which is fatal to all flavour, the victims get dried up to leather. Game, and even poultry, should be drawn as soon as possible after death, but in hot climates the feathers should not come off till the last moment. They prevent evaporation and keep off insects. Of course, all this does not apply to game of which the skins are to be saved as specimens. The sooner the skin is off, the better for this purpose ; but then the carcases had better be used up in soup except with a few coarse birds eaten only for want of better, as "a change on the everlasting mutton and moorghie." Of these are the bald coot, the Brahminy duck and the " beefsteak birds" and ibises (commonly called curlews). Sand grouse ought to be *kept in their skins*, but skinned just before cooking.

To return to our flamingo, he is only found in our present province on a few large tanks and rivers, and does not breed here. It seems to be very uncertain when he *does* breed, but the first flocks fly southerly on the Indus in September, like those of other migrant ducks. The flamingo rarely swims, but will sometimes do so on a tank or river rather than take the trouble of flying from one sand bank to another. On one occasion I shot two of a flock which lit and swam in three fathoms of salt (and rather rough) water on one of the creeks of Bombay harbour. This was on the 28th May, very late for a migrant bird. They are said to run sometimes, but I never saw even a winged flamingo so far forget his dignity. It is probably known to most of my readers that flamingoes shovel up their food with the *upper* mandible, turning the head quite upside down, in the position of the Gordian acrobat, " with his grisly head appearing in the centre of his thighs." I have seen drawings of a variation of the bill of the domestic duck, produced by cultivation and selection, exactly like that of the flamingo. The breed was said to be German, but how these ducks fed was not recorded by my authority. A flock of flamingoes in flight, with the sunlight on their red and white plumage, is a lovely sight. They usually fly in a rather irregular wavering line, the centre birds much higher than the flankers ; and I have heard a flock likened to " a drunken rainbow." The native names are *Rájháns* (or king-goose) and *Rohi*. The latter is so like the name of the Nilgai in Mahratta that I once supposed myself to be going in pursuit of the "blue bull," when my guide was really taking me to a flock of flamingoes.

Real wild geese do not come into the Deccan or Khandesh, as far as I am aware. The "black-backed goose," "comb-duck" or " *nukta*" (*Sarkidiornis melanonotus*) is found more or less (generally less) over the whole region ; but many people consider him rather a duck, and his habits on the water *are* those of a duck, though his flight is that of a goose. This bird may be considered the representative here of the South American Muscovy ducks, which essentially tropical birds have got their Hyperborean name by reason of a funny confusion between "Musk" and Muscovy. They are supposed at certain seasons to have a flavour of musk. The only other bird of these waters having any pretence to goosehood is the well-known ruddy shelldrake called "Brahminy duck" and "Brahminy goose," and by natives all over India " *Chakwa-chakwi*." It really has much of the build and flight of a goose, and seems to me to lead to the true geese from the shelldrakes, as the " *nukta*" does from the ducks. Particularly it has a goose's habit of grazing on young grass or corn, and this makes me ·very unwilling to accept Mr. Hume's charge against it of eating carrion. This idea may have arisen from a mistake between this bird and the similarly coloured Brahminy kite (*Haliastur indus*) caused by the mirage which hangs over the sandbanks that they both haunt. I have myself carefully stalked what I took for a Brahminy duck in the bed of the Tapti,

to find, when within range, that I had wasted my pains on that
" greedy gled." If, however, a carcase of any animal were lying
half in the water, it would attract the crustacea, to which no duck
objects. I do not know any season at which this bird is anything
but a last resource for the pot, but it is sometimes shot for the sake
of its very handsome plumage.

Of the true ducks, the European mallard (*Anas boschas*) is not, to
the best of my belief, found in the Deccan or Khandesh at all. When
any sportsman of those parts tells you he has killed so many
" mallards," he generally means the closely allied spot-billed duck
which is found here, with the shoveller, gadwall, and pin-tailed ducks
and the white-eyed duck (*Aythya nyroca*), which would be far better
named the white-*winged* duck from its white speculum, the colour of
the eye being very far from constant. It is small, and not usually
considered a first-rate duck for the table, but this depends a good deal
upon its diet, which is, I think, a little miscellaneous. I have heard
single specimens highly praised by competent epicures. This bird, the
shoveller, and the blue-winged teal are perhaps the commonest
ducks of the region, and certainly make the longest visit. The
common or grey teal of Europe is also well known here, but
on the whole the country is a bad one for ducks. The mergansers
and the true shelldrake are not found here at all.

Of the next tribe, the grebes, we have one, very common, the
dabchicks, probably identical with the European bird, though some
naturalists separate it. At any rate it is similar in appearance and
habits. The Mahrattas call it " *Pan-buddi*" or " water-diver." It is
a great enemy of fry and spawn ; useless for any human purpose ; but
it gives life, often enough, to waters that show no other swimming bird.
It is sometimes shot as a " teal," a mistake which could not, I should
think, survive the first mouthful, but I have not tried. It is a permanent resident, and breeds in some quiet places.

On large rivers and tanks one occasionally sees the brown-headed
gull, and daily some species of fresh-water terns, very beautiful and
graceful. These eat an enormous quantity of small fish and crustacea,
and moreover forage ashore, chiefly for grasshoppers. I have not
found the nests of any of them in this region, although one might
well expect them to breed on the sandbanks of the larger rivers. The
strange black and white skimmer (*Rhynchops albicollis*), which looks
something like a tern, is not, I think, found here, though it does
exist on the lower waters of our rivers beyond our boundary.

Only one tribe of birds remains to notice—the fishing birds proper, headed by the pelican. I have once seen the great white pelican of Europe in Khandesh, and the Indian grey pelican is occasionally met with all over the region, and may breed in it. The smaller white pelican may be found, but I do not know of any record of it here. Pelicans, indeed, want more fish and bigger fish than they can often find in our present waters. Even their lesser kindred, the European and Chinese cormorants, are not common, probably for the same reason, but another poor relation, the little cormorant, *Pelicanus javanicus* is everywhere. There is hardly so small a puddle that you will not find one or two of these amusing birds on it, and on very moderate-sized pools a flock will alight and worry the water in all directions till every fish, crab, and prawn is either eaten or driven into cover. They have favourite roosting places to which they fly from a long distance, and about sunset the flocks follow each other rapidly, always following the course of the water. They are bold and familiar birds, and will come and fish in front of a tent for hours, and sometimes attach themselves to buffaloes in the water, as cattle-egrets do. A solitary buffalo, which used to spend its day in the water near my tents, was attended by, apparently, a particular cormorant, who would dive off on one side and come up on the other, passing even between the fore or hind legs, and then spreading his wings to dry as he perched the buffalo's head or back ; the ·latter did not seem to object at all. Probably his body attracted small fish, of which some species are very curious, and will come bobbing their noses against any new object, to the great discomfort of nervous or thin-skinned bathers. It is just possible that they know enough about a buffalo to calculate on finding ticks on him, but this is a mere conjecture. The little cormorant is much given to perching on trees. Even the larger European cormorant does so more freely here than in Europe, confirming the statement in Paradise Lost—

> " Upward he flew, and like a cormorant,
> Perched on the tree of life."

Milton can hardly have had many opportunities of observing cormorants ; and I have even known the passage to be criticised by English observers as untrue to the habits of the bird, but the poet was right. The Mahrattas call the cormorants " *Pán-káwala,*" or water-crow—a very good name. This bird breeds in trees, and no doubt sometimes within our region. But I have not got the nest here, and I have noticed that cormorants are scarce in the Deccan in the rains, when the muddy and violent currents are unfavourable to their fishing. I think it likely that most of them migrate to breed; probably to the lowlands of the East Coast. I did once know a man who declared that cormorant soup was very good, but I can't say I have tried it. My friend's pot was supplied with meat of *Pelecanus carbo*, but probably all species of the genus would have much the same flavour, and that a strong one. It would be a good thing if any use could be made of *P. javanicus*, for the ravenous little bird probably diverts more fish from the human dinner table than any other bird or beast except the paddy-bird; and these two together, I think, eat more fish by tale, in this region, than all other bipeds and quadrupeds put together.

The next bird (and the last on my list) can do more as an individual, but he is not nearly so common. This is the "snake-bird" or "darter" (*Plotus melanogaster*), a "cormorant with a heron's head and neck."

This bird may be found on all the deeper streams, but in this part of India not so often on tanks, probably only because the Deccan tanks very often offer no good perching places, or are too much disturbèd by men and cattle, for elsewhere the snake-bird is as apt to be found on a tank as on a river. He delights particularly in wooded streams and in trees that overhang deep water, but I have never seen him plunge from such a position to catch fish like a king-fisher, as an American species is said to do, whence the name "darter." Nor does he fish from the wing, but entirely by diving like a cormorant. His flight, however, is much more lofty, powerful and graceful than that of any cormorant ; and he frequently soars for a considerable distance without apparent motion of the wing, which the larger cormorants can do only to a very limited extent, and the little cormorant not at all. I have never got the nest of this bird, and I doubt his breeding in the Deccan or Khandesh. If he does so, it is probably in the hills, but, as with cormorants, the diminished number of "snake-birds" in the rains makes me think that they emigrate to breed perhaps to the " Bengal side of the punkah," where Dr. Jerdon found them most plentiful. They are much hunted for the beautiful black and white scapular plumes, which have their edges as it were "Italian-ironed." There is no prettier plume for a hat than the bunch from one wing of a snake-bird, with a few white egret feathers set behind it and rising above it. The season for shooting the birds is in the cold weather; some of them begin to moult in April, and by May not one of them has a feather fit to be seen. The moult is often so complete that the bird altogether loses the power of flight, and must remain on a favourite pool for some days. Like all the tribe, it can scarcely move at all on land. It is generally easy to see before firing whether a bird is in good plumage or not. If it is sitting out of the water, or flying, the silvery plumes and similar coloration of the wing are pretty visible, and when it is in the water, showing only the neck and head, or flying overhead, the neck tells an old plume-hunter whether he should spend his shot. In good specimens the neck looks almost white ; in moulting birds it is much darker.

It is a mistake to shoot a snake-bird sitting, as the plumes are likely to be damaged by shot. He should be taken in the water, when he shows only the head and neck, or on the wing from below. In the former case small shot should be used, as the thin neck forms a very narrow target.

Of fresh-water reptiles we have in the Deccan region, first of all, certain water tortoises or terrapins, easily distinguished from land tortoises by their webbed feet, and from the fresh-water turtles by their " tortoise-shell" back and breast-plates, and by having either five or four visible claws on the fore feet and always four on the hind feet. Curiously enough, while the American terrapins are of most delicate flavour, ours are uneatable, smelling foully, as is indicated by their untranslatable Mahratta name. They are carnivorous, and are sometimes caught on a live bait, or on a worm, or bit of raw meat. Some that I kept in confinement refused carrion. The natives often put them in wells, especially *Emys trijuga*, the commonest species, and call them, as well as all other tortoises, and turtles, " *Kásaw*." All " Kásaws " are supposed to be poor relations of the great turtle, who upholds the world, and are accordingly respected by the more pious Hindus, and an image of a tortoise is often to be seen on the floor of a temple. This has something to say to the putting of them in the wells, but they are useful there as scavengers, and as mortal enemies of the fresh-water crabs (*Telphusidæ*), which do a great deal of harm to wells by burrowing in the foundations. They cannot, I think, do much in the way of catching live fish, for I have known them to be in wells with fish for many months without any diminution in the number of the latter, though there was apparently no other food. Probably frogs, crabs, mollusca, and insects form their chief diet ; and it may be, as I shall show reason for believing with regard to the next group, that they have been too hastily pronounced "exclusively carnivorous."

This next group is that of fresh-water turtles.

These are, compared to the terrapine, very flat and round, with a distinct edge, something the shape of two saucers put "lip to lip." They don't show any " tortoise-shell" at all, but a smooth leathery surface, flexible round the edges. In front and behind, this flexible edge is double, and obeys the voluntary action of the muscles, at least in young specimens, which, after drawing their heads within the shell, will close the edges of the upper and lower leathery flaps till they almost touch each other. These fresh-water turtles have all been classed as carniverous, though Dr. Kelaart long ago recorded that one (*Emyda ceylonensis*) in his possession fed freely on bread and boiled rice. I have repeatedly myself taken wild specimens with paste baits, and have seen them assemble under a wild fig tree (*Ficus glomerata*, the *Umbar* or *Guler*), of which the ripe fruit were dropping into the water, and apparently taking the figs. It is true that a ripe wild fig is usually so full of maggots that it constitutes a "mixed diet."

In the courtyard of the Black Mosque of Ahmadnagar, long ago desecrated and now used as a public office, there was in my day in a small cistern a fresh-water turtle, about 18 inches long, who had been there as long as any one could remember, and is probably there yet. The water was filtered, and the feed-pipe grated, and so little food would have come to him by that road, and to put any kind of animal food in the cistern would have polluted the water for many people and caused trouble. The turtle was regularly fed by his neighbours with vegetable food, especially, in their season, with parched heads of maize, which he was very fond of. Specimens in my own possession were fed on fresh dead fish, and refused carrion.

They are often taken by the angler with live bait, or raw meat, or worms, and sometimes, as already mentioned, with paste. They give more fun sometimes than one would look for, but often cut the line with their gouge-like jaws, or get into a hole, or bury themselves in the mud ; and often when landed, it is found that they have gorged the hook and the trace must be cut, and the hook recovered afterwards by the cook. It is necessary to use great care in handling them, as they bite savagely, and can take the piece out ; the jaws are like two gouges closing on each other.

They make very good soup and curry, and I have been very much amused at a friend's refusing the former when he knew what it was, who had probably often enjoyed the like before, under the belief that it was made of a sea-turtle. They are put into wells and cisterns in the same way as the terrapins, and for the same reasons *Trionyx javanicus* is our commonest species, and *Chitra indica* the largest. I have seen a bullet glance off the shell of the latter, but it was fired at a considerable angle. The turtle was afterwards killed by another bullet, fired almost vertically down upon the centre of the back, which passed completely through him. These fresh-water tortoises and turtles, if *turned* on their backs, speedily recover their proper position, using their long necks and heads in doing so.

The crocodile (Mahratta "*magar*," "*suswar*") is only locally common in this area, very seldom seen in the tanks and smaller rivers, but occupying particular deep reaches in the great rivers, often in considerable numbers. These are the places to which the larger fish and the turtles (crocodiles are very fond of turtles) retire when the rivers shrink in the dry weather, and where, accordingly, food is plentiful. As far as I am aware, there is only one species known here, viz., *Crocodilus palustris.* I have measured specimens from the Upper Tapti and Bhima 10 feet long, and I do not think that that size is often exceeded here. And though I have heard many crocodile yarns, I do not myself know a single well-authenticated instance of a crocodile's killing a human being in the Deccan of Khandesh. Once, in 1875, I remarked as much to a native official, who immediately said that a man had been killed by one in his "Taluka" (or barony) "last year." Being asked for details, he gave them, upon which I recognised the story as one I had heard in the same place in 1872 as of "last year." I dare say that crocodile is killing that man "last year" to this day. The other form of crocodile-saga always refers to the "next village," and when you get *there*, to the next, and so on, slipping away before the inquirer like the foot of a rainbow before the infant gold-seeker. I believe that the larger and more dangerous *Crocodilus porosus* is found in the lower waters of most of the great Deccan rivers beyond our boundary. The differences, setting aside size and temper, are that *C. palustris* has two sets of shields on the back of his neck, arranged in two groups of four and six respectively (the four in front), six shields in each transverse row of the middle of the back, and sixteen such rows of dorsal shields altogether to the root of the tail. But in *Crocodilus porosus* the "anterior nuchal plates" are none, or only 2, and then rudimentary, that is, his cousin has a front set of 4 plates on the back of his neck; and he has not, or only two little ones. Its dorsal shields are usually six in a row on one part of the back and eight in the rest (the extra two rudimentary), and there are 17 rows in all, to the root of the tail.

I need hardly say that alligators are not found here, nor anywhere else in Asia, except China, where there is one rare species. The outward and visible sign of a crocodile proper, as distinguished from an alligator, is the fourth tooth of the lower jaw on each side, which grins alike at all seasons, whether the mouth be shut or open, improving a naturally ugly countenance with a hideous fixed snarl. In the alligators, this tooth is received into a sort of sheath or pit in the upper jaw. Some alligators, moreover, have shields on the belly as well as on the back. I have wasted a great deal of time on catching crocodiles, and never caught one, though others have had better luck. Shooting them with the rifle is really good sport. This should be done in the heat of the day, when they lie on banks in the sun. In the morning they are wideawake, and before sunset they begin to forage. They have to be carefully stalked and clean killed, otherwise they get away into some hole, or (I think) bury themselves in the mud, as they are well known to do sometimes, in lakes that dry up for a season, to await the return of the water. Many a hit crocodile goes off leaving a trail of blood on the water, and is never seen again. But if they remain in one spot even for a few seconds after receiving the bullet, that is a sign that they are very hard hit; and in such a case the carcase will generally float within from 30 to 40 hours. I have not had a harpoon that could penetrate the back scales; a good hog-spear, however, does so easily.

The story of their being ball-proof arises chiefly, I think, from the natural unwillingness of man to admit that he has missed. A very ordinary gun will put a bullet through and through any part of them, unless, perhaps, the bullet strike at a very great angle and glance off. I believe that this once happened to a bullet fired by myself from a very light fowling piece. A shot in the small of the back head, heart or spine will stop them easily enough. Behind the shoulder is the best shot from the side; but if you shoot from above, as from a high bank or a ship, aim at the root of the neck. Not only is it a good place, but the places above and below are good too, and the usual error of a rifle shot is high or low.

A crocodile, lying on a bank, covers his heart (to a great extent) with his left elbow, and a light express bullet will break upon the bones of the arm, doing little hurt. When struck or startled in the water, they will sometimes leap forwards, three or four feet from the surface, like a salmon, and once I saw one, shot through the heart on shore, literally stand on the end of his tail for a second, and fall backwards stone dead. They are not heavy animals; the largest I ever weighed, a female, 8 feet long, was only 100 lbs. in weight, though full of eggs. They are not of much use when you have got them. The bleached skull makes a ghastly trophy, and the skin a very ugly one; but I once got two very handsome shields made of crocodile skins at Ahmedabad. Here I may remark that I have never got the traditional *bangles* from the stomach of any crocodile. I have got sticks; what the brute ate them for I can't imagine. The handsome leather used in Europe for cigar cases, bags, and so forth is all made of the skins of young American alligators; the art has not found its way here yet. Natives use the teeth and shields for charms and the oil for medicine, and some low castes eat the flesh and eggs. There used to be a small tribe in the Tapti valley who devoted their lives to hunting crocodiles, and showed great pluck and skill in it. They used nets, nooses and broad-bladed pikes (not harpoons), and always cut the tail with an axe as soon as possible,—à trick known to other natives besides them. Crocodiles are commonly supposed only to *crawl*, but the young of *C. palustris* can walk and even run. A recent observer has noted the same in Ceylon. I have twice kept young crocodiles alive; they were savage and sulky, refused food, and threw it up when administered by force.

Of other water lizards we have only *Varanus dracœna*, the *Ghorpur*, which, however, chiefly comes under notice when *out* of the water, of which it is very independent. It is lucky that Ghorpurs don't get to be much more than four feet long, for they are very active and greedy, and I have seen one much shorter than that wage a good fight with a small terrier dog. They will eat any animal that they can overpower and swallow, up to young ducks, and I have no doubt that they would eat the old ducks, too, if they could either swallow them whole or carve them in any fashion. They destroy eggs of all sorts, but I don't quite understand how. They don't swallow them whole, for the shells are left.

Young Ghorpurs are among the various lizards, supposed to be venomous and called "*Biscobra*" in this region. The Biscobra of Sind is an Eublepharis, according to Mr. Murray, an ugly creature certainly, and looking really very like the known venomous Heloderma of South America. Mr. Murray found the *secretions of its skin* really to some extent poisonous.

This is no place for going into so long a list as that of the fresh-water snakes. It is, perhaps, enough to say that, although almost all snakes swim well, only those to the manner born can dive well, and it is easy enough to tell the difference between a true water-snake and a mere passenger by water. The latter holds his head much higher, and never stays still in the water, but "keeps moving."

Some snakes, however, are amphibious, and one of these (*Tropidonotus quincunciatus*), the spotted water-snake, is very much commoner here than any of the true fresh-water snakes. They are sometimes caught on hooks, when a frog or fish is the bait, and then they foul the tackle, and make the angler unnecessarily nervous. None of them are poisonous, and I do not think that any venomous land snake is sufficiently at home in the water to take a bait below the surface. This tropidonotus is the "*pán-diwar*" of the Mahrattas. There are several varieties of colour. Those in dark, muddy, shady waters are a sort of dull tortoise-shell colour; and some in open tanks and streams might almost be described as black and gold. There is one very libellous sort of snake-story which describes water snakes as climbing up boat's cables to bite people on board. Now, a fresh-water snake could have no motive for going aboard at all; and if he did go aboard and bite people, they need no more die of it than if he was a mouse. As for the sea snakes, which *are* all venomous, they can hardly crawl on the sand, let alone climbing up a cable. But no doubt a really poisonous land snake, swimming across a river, might think a boat a good place to rest in. A cobra or bungarus would easily enough get up the cable, and his misdeeds, if any, would be laid upon the innocent water snakes. Probably, however, most accidents of this sort arise from snakes being brought on board in cargo or firewood.

Of frogs (*Menduk, Bhenki*), we have many. The most conspicuous is the big bull-frog (*Rana tigrina*), an unpopular creature. He eats pretty nearly whatever creature he can catch, and *vice versa*; reminding one of the ancient Gaelic proverb, "This is the government of the waters; the beast that is greatest eats that which is least and the beast that is least shifts for itself."

The next and less known is *Rana esculenta*, the very identical French frog. For want of French cooks he is wasted here upon the

storks and catfishes. I never saw *Cacopus globulosus*, a marvellous
frog figured by Dr. Gunther, the very representative of Humpty
Dumpty among reptiles.

Natives don't usually pay much attention to frogs; but once when I
had a lot of men stung by scorpions, a village elder made cataplasms
of live frogs pounded between stones, and applied the quivering and
mangled reptiles to the injured parts with great success. I think the
very nastiness of the remedy gave the sepoys more faith in it.

Tigers are said to eat bull-frogs in the rains, and thereafter to
sicken and waste away, just as in Ireland a skinny cat is supposed
to have been eating crickets. I think myself that the tiger is probably pretty far gone in famine before he takes to catching frogs,
and it is pretty certain that all the frogs he could catch in a day
would make him but a poor day's ration.

Of the crustacea of our fresh waters we know but little, and have
no standard books on the subject. Crabs (*Telphusida*) are found
almost to the top of the ghâts, and furnish food to man, birds, turtles
and fishes. They are said to be unwholesome in the hot weather,
which is not borne out by my own experience. And at that season
certain forest tribes go and grind stones on each other in dry nullas.
They say that the crabs mistake the noise for that of waters. At any
rate the crabs do come out, and are caught and eaten. Another plan
is to drop a bullet or pebble, attached to a string, into the crab's hole,
who thereupon nips it and is drawn out holding on to what he, no
doubt, supposes a live intruder. The Mahratta names for them are
Kenkad and *Muta*. The former word, with a dry humour characteristic of that nation, is also applied to *handcuffs*. I have good
precedent for introducing these useful articles into my paper, for the
United States Commissioners to the Fisheries Exhibition exhibited a
pair with a label stating that they were found " very serviceable in the
whale fishery; and carried by most vessels."

A true prawn is found even above the falls of the Godavery, and
small shrimps up to at least 2,500 feet on the ghâts. These latter are
sufficiently abundant to be dried for sale. A cray fish in the streams
of the Satpura is said to reach " a cubit" (*hát*, 19 inches) in length
over all, and fragments that I found bore out the statement. I use
the term cray fish here, as it always has been used in English and
French (*ecrevisse*) to mean a crustacean *with* nippers. Some
naturalists have attempted to restrict it to those that have none, but
the limitation is artificial and cannot succeed.

Of mollusks, the most part are water snails, the most noticeable being the great round ampullaria, as big as a baby's fist. There are at least two mussels (Unio), one with a rather delicate shell and pale olive green epidermis, and one far more solid and of a black or dark brown colour. The latter is said with great probability occasionally to contain pearls. One such pearl is in the Kolapur Museum, and some from Bengal were exhibited at the Fisheries Exhibition. The natives call all univalves *Kuba* or *Kubi*, and all bivalves *Shipi*, or some derivative thereof.

Insects of all sorts swarm in and near the water, but there is no space here for describing them. I do not think that anything like the appearance in swarms of the European Ephemerides (green drake and May-fly) is ever seen in this region. The nearest thing to it is when a swarm of newly-hatched winged white-ants drifts over a river or tank, when the fish may be seen rising at them all over the surface. The same thing happens, but more rarely, with locusts. Mole-crickets, wherever obtainable, are a very good bait for almost all sorts of fish. Waterbeetles attain an enormous size, and no doubt destroy fish spawn and even small fry.

Earthworms (Mahratta *gándúl* and *gandrín*) are generally very hard to get, but when they can be got are as useful here as at home. Leeches (*Jalu*) are sufficiently numerous in some tanks to make bathing impossible, but are not otherwise a plague as in some other tropical countries.

A notice of these waters would hardly be complete without some reference to the daily visits of terrestrial animals and birds to the water which are always a remarkable feature of animal life in dry hot climates. The large carnivora usually drink just about dark, perhaps a little before or after. It is said, too, that after eating they always go to the water, at whatever hour. The small cats do the same; but the jackal usually drinks about 9 or 10 A.M.; and the mongoose and civets even later.

The larger wild ruminants, where much hunted, drink before sunrise and after sunset; but when undisturbed, or after any specially thirsty business, such as love or war, will visit the water at broad noon and before sunset. The small four-horned antelope and the barking deer prefer noon-day; the gazelle usually drinks a little earlier, say, 10 or 11 a.m.

But the general drinking time for birds and beasts is when the morning begins to warm up, say, from half-past eight to half-past

nine or ten a.m., when all diurnal animals have been abroad all
morning, and want to wet their throats before retiring, probably
to keep quiet for the day. The grey partridges and francolins are
amongst the first to steal down to the water ; and after them come
the common sand-grouse ; pretty common in this region. The
painted grouse, which is found in low thorny jungle, is an exception.
It drinks by twilight, often so late that it is only recognised
by its very peculiar chuckling note. But after the common grouse
(*Pterocles exustus*) come, if there are any about, the pea-fowl, blue
pigeon and doves, more rarely the green pigeon (*Crocopus chlorigaster*), according to Dr. Jerdon. I have not myself seen this bird
drink, and one I kept in confinement did not seem to care about
water, getting much moisture in his juicy food. The authority
however, is conclusive, and I have myself noticed the green pigeon
to be commonly found in trees near water about 9 o'clock a.m. and
a little before sunset. The monkeys also drink at this hour (9 or 10
a.m.), and so do crows, who take a regular bath, with a good deal
of demonstration, as in all their doings. Eagles and hawks come about
the same time, and sometimes stand in the water, apparently merely
to cool their toes. When any of them look out for fish or frogs they
do it on the wing.*

Pretty much the same thing happens again from about an hour
before sunset to half-an-hour after it ; but besides this the water,
if there are any trees or bushes near it, has always a tendency to
become the centre of all animal life ; and the angler, perhaps, sees
more of this than he would if shooting, or even walking, and for many
reasons it is well that he should have a gun-bearer at hand.

This is hardly the place for discussing fishing-nets, but the best
to have in a camp is the casting-net, which can be handled by one
man. If you have two fishermen, this may well be supplemented
by a *gholni*, or shove-net, fixed to two bamboos, and with a large
party a seine can be used generally ; wherever the water is large
enough for the use of a seine, native fishermen will be found in possession of one, or will improvise it by linking smaller nets together.
A small boat is useful in " shooting" the seine ; and the best portable
boats are certainly the canvas " Berthon boats." It is not, however,
easy to shoot from them unless after carefully ballasting them, or
fitting an outrigger ; for, although very hard to upset, they are very

* NOTE.—Several eagles, especially the serpent eagle (*Circaetus gallicus*), catch
frogs on the marshy borders of tanks.

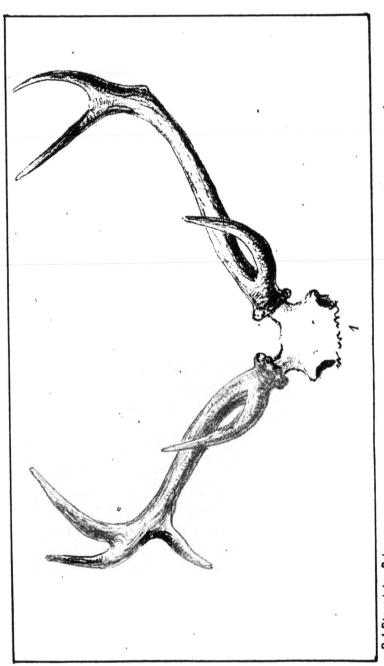

R.A. Sterndale. Del.

easily made to rock, and even the putting up of a gun to the shoulder will cause enough motion to spoil the shot. The same is the case with small native canoes, and the remedies are the same. Safe, though clumsy, rafts are made of gourds lashed to a charpoy or of bull-rushes by the natives, but these are apt to sink a few inches below the surface, and should be surmounted by a bath-tub, a pair of wine boxes caulked and painted, or some similar device for keeping the passenger and his ammunition dry.

In some places the natives make round coracles of hides; and in others they use huge circular sheet iron sugar boiler for boats; in either case reminding one of the Wise Men of Gotham in their Bowl.

KESWAL.

ON ABNORMALITIES IN THE HORNS OF RUMINANTS.

By R. A. STERNDALE, F.Z.S., &c.

There being several striking examples of deformity in the horns in the Society's collection, I am induced to bring them to notice and to theorize on the causes which have led to such results; and a varied field for speculation is opened, for many questions arise in connection with the subject. The first is, are these abnormalities, in the case of antlered ruminants, transitory or persistent? and, secondly, in the hollow-horned ruminants is the *fons et origo malis* in the osseous or horny formation? Then comes enquiry into the primary cause of such malformation. The whole subject is involved in doubt, and but a mere hypothesis can be arrived at, for almost every day we come across some freak of nature which starts us off into a new channel of conjecture. With regard to the first question, are the deformities of deer transitory or persistent? that is to say, would a Sambar Stag, who had developed in his seventh year an abnormal tine, reproduce that abnormality the following year—the eighth? or would he revert to his normal form? Now I will give an example from a very fine head in my own collection: the horns are unusually large, the right beam being 45 inches and the left 43 inches in length; on referring to figure 1 in the accompanying plates you will observe a tine of 9 inches long, which is a decided abnormality; there is no reversion or progression towards lower or higher types, but simply a sprout which has taken a direction quite out of the symmetry of known species. Now, to arrive at any conclusion one must consider the process of the growth of antlers: they are produced annually, and with a tendency to increase instead of decrease; on the shedding of the old horn there is a decided determination of blood to the head in the animal ; the new growth, a fibro-cartilaginous substance, is nourished by blood vessels, which ramify on the exterior, covered by a sensitive velvety skin ; whether this be true venous blood or a specialized fluid of a more albuminous nature is a question which has not as yet to my knowledge been solved. Anyhow, a blood-like fluid is conveyed along the growth of the horn feeding the bony deposits, and it may be that abnormal sprouts are the result of an aneurism in one of the blood channels ; but if this be so, my horn brings up another question, for if you will look at the normal antler you will see an excrescence exactly corresponding with the extra tine, yet not so fully developed. Is this the sympathy that one sees exemplified in cases of toothache ? The decay of a particular tooth on one side is frequently followed by that of the corresponding one on the other. If this particular stag had been allowed to live for another year, would both antlers have shown an additional tine, or would they have reverted to the normal shape ? There is no reason why such deviations should be perpetuated in the same individual or transmitted to his descendants. It was thought at one time that the spike buck of America, which is the many-antlered *Cariacus virginianus*, found occasionally with a single-spiked horn, was a freak of nature transmitted from the first so formed buck to his progeny, and this was gravely advanced in an American Scientific Journal, and it was asserted that the spike horn bucks were gradually crowding out the antlered ones on the principle of the survival of the fittest, however better informed naturalists like Judge Caton proved that these were merely young bucks of the first year whose second season saw them with branch-ing horns.

I am inclined to think that there is neither persistence nor transmission in the abnormalities of antlered deer. I believe in injury being the cause of these freaks.

Sympathy in certain cases of bodily injury affects the horn of that particular side, and this is permanent through life, and in such cases the horns are not shed.

There is a curious bifurcation of the tip of the bez tine in the right antler of a Cashmere Stag's horns in my collection, which must have occurred whilst the point was tender ; and this reminds me of what I have recently read in the second volume of the transactions of the Linnean Society of New York regarding the growth of antlers. It is the com-monly received idea, accepted by most naturalists, that the blood vessels contract at the burr or base of the horn on its arrival at full growth, and that then, the velvet dries up and is rubbed off by the animal, but the Hon'ble Judge Caton, of Ottawa, Illinois, from observations made in his own deer park, states : " The evidence derived from a very great

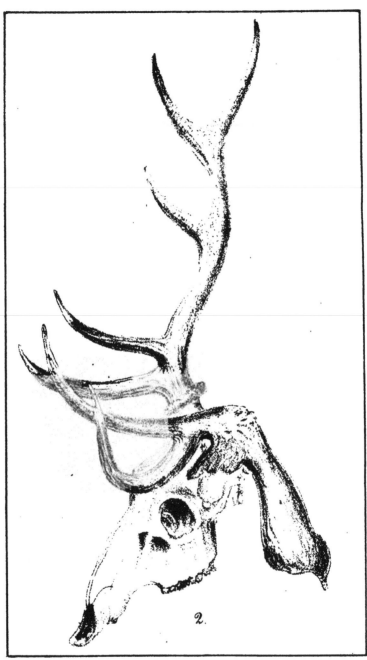

2.

R.A.Sterndale. Del.

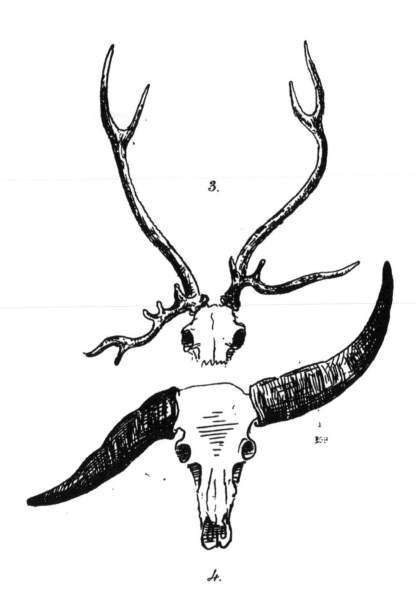

3.

4.

multitude of observations, made through a course of years, is conclusive that nature prompts the animal to denude its antlers of their covering at a certain period of its growth while yet the blood has as free access to that covering as it ever had."

It is the common impression that the animal is extremely sensitive to pain whilst the velvet is in its quick state. I am, however, informed by Mr. Phipson that he has seen the old Wapiti Stag, we most of us remember, near the entrance gate in the London Zoo, rubbing his huge antlers whilst the blood flowed freely from each abrasion.

Now I come to a very curious deformity in the Society's collection— figure No. 2. It is that of the left antler of a Cashmere Stag ; the right antler is perfectly symmetrical, but the left one, as you will observe, is broken and bent down about 2 inches above the bez antler, and instead of branching it has formed itself into a club. There is no doubt of the fracture here—it is self-evident. Either from a fall, or a blow from a falling branch, or from some such injury the soft antler was broken, but the velvet held on, and the nourishment continued, but in an interrupted way ; the free circulation, was impeded, and instead of the tines branching out according to their wont, they coalesced into a knob as we see it here. Of all the deer tribe, I have found the Axis. or Spotted deer most given to "sports" in its horns. The normal shape is strictly rusine with three tines, yet 20 per cent. of horns show little sprouts generally at the base of the brow antler. Figure 3 represents one in the Inverarity collection, in which the brow antlers have run riot altogether and the right one has thrown out several branchlets. Probably in this deer there was something constitutionally wrong. I have examined all the deer in the Victoria Gardens and have noticed in the largest stag in the Axis pen, which has very fair-sized horns, that each brow antler has an abnormal branch.

Though it is thus easy to build up a theory on the deformities of the antlered ruminarts and to speculate on their persistence, a new train of thought aris entirely in connection with the hollow-horned ruminants. In these abnormalities must be persistent ; with them it is an exemplification of the adage "as the twig is bent, so is the tree inclined," and as their horns are to a certain extent supported by bony cores, it is in these we must look, in the first instance, for the deviation from the usual symmetry. Figure 4 represents a buffalo head, the property of Mr. Inverarity, at present deposited with the Society ; the deformity here clearly begins with the bony core ; with such soft and easily deflective material as horn eccentric shapes can be artificially produced, but the deflections must be beyond the limit of the bony core ; in the case of this buffalo the deformity, or rather wrong direction, begins from the base and must have been regulated by the core. It is not an uncommon thing to find antelope horns running up almost parallel to each other instead of the usual V shape. I have two such in my own collection. Here the core again gives the direction, and in the numerous cases reported in the *Asian*, and elsewhere, of antelope with distorted horns the core is evidently the source of the eccentricity. Figure 5 gives a sketch of an antelope head in the Society's collection ; the deflection starts from the base, and the bony core is evidently so twisted that I have not been able to unscrew the horn as can usually be done with dried antelope heads. The horns of tame buffaloes frequently show deviations from the normal type. There is in Bombay at the present moment a magnificent old buffalo with grand horns of a most curious and perfectly symmetrical shape. They are very massive, and come down low, close on to each cheek, and then sweeping round with a curve form a perfect circle at the tips.

A LIST OF THE BOMBAY BUTTERFLIES IN THE SOCIETY'S COLLECTION,

WITH NOTES BY MR. E. H. AITKEN.

The butterflies in the Museum of the Bombay Natural History Society are geographically divided into the following collections :—

(1.) A fairly representative, though by no means complete, collection from the Bombay Presidency, exclusive of Sind on the one hand, and Canara on the other, which latter belongs rather to the Malabar region. For these the Society is indebted largely to Mr. R. C. Wroughton, also to Mr. Moscardi, C.S., and other members. This collection is arranged and named.

(2.) A very incomplete collection from Malabar and Canara, partly purchased and partly contributed by Captain T. M. Macpherson. These are arranged and partly named.

(3.) A small collection of British butterflies presented by Mr. R. C. Wroughton.

(4.) A small collection from different parts of the Himalayas, partly obtained by exchange and partly contributed by members.

(5.) A few, interesting, named specimens from the Punjab and from Aden. These were the gift of Major Yerbury.

I take more interest in butterflies on the wing than on the pin, but that the following notes may serve a double purpose I have based them on a list of the species in the first of the collections enumerated above. I named the collection myself, so that no one else is responsible for the accuracy of the list, and I must protect myself at the outset by disclaiming any pretence to give a complete or discriminative catalogue of the collection.

5.

R.A Sterndale. Del.

In the present unsettled state of the subject it would be impossible to attempt such a thing without diverting a great deal more of my leisure than I am willing to divert from nature to nomenclature, and I am besides peculiarly disqualified for such a task by my inability to believe in a great many of the species which are accepted by those who seem to be pillars. This will account for the absence from my list of a good many species, under one or two genera in particular, such as *Terias* and *Teracolus*, which, if they are species at all, are very common.

I have no systematic notes of the months in which I have caught each species. I regret this, but at the same time I think that the data obtained in this way may be over-valued. Suppose from such notes you deduce the fact that *D. chrysippus*, for example, may be met with every month in the year, is the fact worth recording? There is no butterfly which may not be met with any month in the year, for some pupæ always remain over from one season to the next, and an accident may bring these out at any time. What we want to know is when each species is in season and why? Almost every species has a well-defined season, depending on its food plant. For the great majority this is the latter half of the monsoon, and the two months following, *i.e.*, the period during which the annual vegetation called into life by the rain remains green. Another season is the commencement of spring, which even in this country makes its influence distinctly felt. *A. viola* comes out at this time. Some species appear at neither of these seasons except by accident. *Virachola isocrates*, for example, where it feeds on the pomegranate, can only be in season when that fruit is ripening. I have tried, as far as I can, from memory and notes, to give the limits of the time during which each species is in season.

NYMPHALIDÆ.

DANAINÆ.

1. *Danais chrysippus.*—This, with the exception, perhaps, of *Terias hecabe*, is the commonest and most ubiquitous butterfly on this side of India. At Kharaghora, on the edge of the Runn of Cutch, this was one of the very few flying things I could get, and my chameleon would starve rather than eat it. I never found the larva on anything else than *Calotropis gigantea*. Dwarf specimens of this are not uncommon. All our *Danainæ* are on the wing chiefly from about August till the end of the year.

2. *D. dorippus.*—There is one specimen in the collection without locality. I have never met with it, but have known of at least one specimen being caught in Bombay. I believe it to be an occasional variety of *chrysippus*.

3. *D. genutia.*—This is common almost everywhere, though by no means so abundant as the last. One specimen in the Society's collection has that

dash of white on the hind wings which is common in specimens of *chrysippus* from Kurrachee (Moore's *D. alcippoides*) and of *dorippus* from Aden. The collection contains also a very remarkable specimen caught at Matheran by Mr. Moscardi in December, 1884, in which the ground-colour throughout is a dull lavender. The markings are normal.

4. *D. limniace.*—This is common too, especially on the hills. I found the larva at Lanowlie in October, feeding cn *Hoya viridiflora.* The offensive smell which makes reptiles and birds—if birds eat butterflies at all—reject th's family, is particularly strong in this species, and is certainly connected with the extrusion of the yellow plumes. It is also a very difficult insect to kill. Pinching the thorax has a temporary effect, but it soon revives. Even when killed past reviving and pinned, it will continue to wag his head and antennæ satirically for some days. This or any of the last will serve very well to illustrate the intimate connection which there is between colour and habit, not where the protection of the insect, by mimicry or otherwise, is concerned, but simply from an æsthetic point of view. On the underside the greater part of the forewing differs from the hindwing, but a well-defined area at the apex is of the same shade. Now in the *Danais* attitude of rest the forewings drop between the hindwings until precisely this portion and no more projects and is visible. For those who like to theorise I would suggest that the action of light has produced this effect, the warmer tint of the covered portion of the forewing representing the original unbleached colour of the butterfly countless generations ago. A *Khakee* coat often illustrates the same thing !

5. *D. grammica.*—This is very common on the hills, but comparatively rare in Bombay. It comes out a little later than the foregoing species, being very abundant about Christmas time. I found the larva at Lanowlie in October last year, feeding on *Tylophora carnosa,* also one of the *Asclepiadeaceæ.* It was, I think, the most beautiful larva I have seen. The ground-colour was a rich reddish brown, or claret colour, and on each segment there was a pair of round yellow spots with numerous small bluish-white spots between. On the sides these spots gathered into a conspicuous longitudinal band. The under surface was black. There were only two pairs of filaments, which were nearly straight.

6. *Euplœa core.*—In Bombay this feeds on oleander, but on the hills I have found the larvæ on the wild fig, *Ficus glomerata.* The larva, like those of all the *Danainæ,* rests on the underside of the leaf, a position which exposes it to the notice of birds; but it affects no concealment, and is evidently not edible. The pupa, like a nugget of burnished silver, seems designed to attract attention. Perhaps it acts on the superstition of its enemies. The natural feeling which forms the basis

of superstition is not confined to us, lords of creation, and I am disposed to think many insects save their lives by availing themselves of it. This butterfly is a great traveller, as indeed are all the *Danainæ*. They are often to be seen crossing Bombay Harbour from one island to another, and it is a curious question whether they see the land in the distance, or go in the spirit of Columbus.

SATYRINÆ.

7. *Melanitis leda.*—This and the next are insects of the dusk, coming out after the sun is down and dancing round the roots of trees in company after the manner of fairies. A little later they come out of their hunts and fly straight up into the sky as far as eye can follow them, for what purpose I cannot guess. They are thirsty creatures, and will gather in numbers where water has been spilt on the ground; but they prefer whiskey. I have found the larva of this feeding on grass. It is difficult to find, being a night feeder and very shy. As the species of grass on which it feeds grows during the monsoon only, except where there is water, this species is in season all the latter part of the rainy season, and in some places for a short they almost jostle each other for room. About October, when vegetation is drying up, it gives place to the next.

8. *M. ismene.*—This is very similar to the last in its habits, and quite as common, more so on the hills. I am aware that they are supposed to be one species, but on this point I have not given in yet. I have noticed it on alighting fall over on one side until it was almost horizontal, which very much enhanced its likeness to a dead leaf.

9. *Lethe neelgherriensis.*—In the month of March this is very plentiful on the ghâts, but it is not confined to them. I have caught it in the neighbourhood of Bombay. It is similar in habits to the last two.

10. *Lethe europa.*—There is only one specimen of this in the collection, and nothing to show where it came from. I have not met with it.

11. *Mycalesis perseus.*—I have nothing to note about this species. I have caught it in Bombay and elsewhere, but it is not common.

12. *Ypthima philomela.*—This is a humble butterfly, flying along the ground in shady places, but it is not specially crepuscular. It is common in the cold season at Poona, and I think on the hills everywhere.

ACRÆINÆ.

13. *Telchinia violæ.*—This is not very common, but a few appear just before the hot season in Bombay and wherever I have been. I met with some at Mahableshwar last March. It seems generally to be on a journey going steadily in one direction with a feeble flight, but it will stop to sip a flower and is easily caught. I believe it is, like the *Danainæ*, offensive to birds and reptiles.

NYMPHALINÆ.

14. *Atella phalanta.*—This is not rare in Bombay, and one of the commonest species on the hills in March, when people go up for the hot season. I imagine it comes out after the monsoon and continues all through the cold weather. It does not remain so long on the wing in Bombay ; but many species have their season later on the hills than on the plains. The larva of *A. phalanta* feeds on *Flacourtia montana,* and is easily found if one knows to look for it, not on the higher branches of the trees, but on the young shoots which come up from the roots. The pupa is a lovely object. So is the butterfly when fresh and iridescent. It is one of the most sprightly and characteristic inhabitants of our hill stations, flitting everywhere from bush to bush and even when it settles moving its wings for ever in the restless way peculiar to it.

15. *Argynnis niphe.*—Colonel Swinhoe, in his paper on the *Lepidoptera* of Bombay and the Deccan, published in the proceedings of the Zoological Society of London, Feb. 18, 1885, says that he caught this in Bombay in 1877. This is very interesting. I have hunted butterflies for years in Bombay and never saw a specimen of this. I can hardly believe that such a conspicuous insect could have escaped me entirely. But looking over the list I find several other species, of which I am equally positive that they are not Bombay butterflies, recorded from Bombay in that year, *e.g.,* *Colias fieldii* and *Teracolus danæ.* The inference is that during the famine year many butterflies wandered, as we know birds did, into regions where they were unknown before. There are specimens of *A. niphe* in the Society's collection, contributed by Mr. Newnham from Cutch.

16. *Pyrameis cardui.*—In Bombay this species breaks out in large numbers at irregular seasons in a way for which I cannot account. It feeds on different species of *blumea,* which are all monsoon annuals, and might be expected to be very regular in its appearance. The larvæ are sociable when very young, half a dozen chumming together under the shelter of a little network of silk. The butterfly is not very easy to catch, being a strong flier and wary. It rarely settles except on the ground, and opens its wings much less than the *Junonias.*

17. *Junonia lemonias.*—Though not rare anywhere, this and the next two are pre-eminently Bombay butterflies, loving its ditches and well-watered gardens. *Orithyia* and *hierta,* on the other hand, like dry situations. In habits, otherwise, they are very much alike, flitting about one spot and basking in the sun all the hottest hours of the day. This species is in season at the close of the rains.

18. *J. asterie.*—Next to *T. hecabe* and *D. chrysippus,* this is the commonest butterfly in Bombay at the close of the rains and for some time after. It attains in old age to a degree of disreputability and raggedness not often seen in any other species. I am inclined to think this is the result of ineffectual attempts to catch it on the part of lizards, with which it is a favourite food. The larva feeds on *Lippia nodiflora* and *Asteracantha longifolia*, both very abundant in Bombay during the monsoon, by the side of, or actually in, water. The larva is scarcely, if at all, distinguishable from that of the next species and very like that of *P. cardui*.

19. *J. almana.*—This comes out at the same season, but is not so common as the last. The larva feeds on *A. longifolia* : I never found it on *L. nodiflora*. Colonel Swinhoe, in the paper above mentioned, suggests that this and the last are one species. I believe the suggestion was made by Mr. de Nicéville before, and the opinion of two such authorities is entitled to respect, but as Colonel Swinhoe appears to quote me in support of his view, I ought to say that I do not share it. It is true, as he says, that I reared both species from a lot of larvæ taken together, but they were taken from a ditch in which there may have been the offspring of fifty parents. This proves nothing. Colonel Swinhoe further says that he has a large series of examples showing every stage of variety between the two. I am disposed to think he might apply the same test with disastrous effect to a score or so of the species which appear in his own list under the genera *Ixias*, *Teracolus* and *Terias* ; but that is a point on which opinions will differ. In this case, at any rate, I doubt the applicabil ty of the test. I have not seen many specimens from other parts of India, but I have reared and caught plenty in Bombay, and I have no hesitation in asserting that here both forms are remarkable for their freedom from variation. For this reason I put down one or two intermediate specimens which I have seen as hybrids. In the Society's collection there is one specimen intermediate between *J. asterie* and *J. lemonias*, which, for the same reason, I believe to be a hybrid, though *lemonias* is a much more variable insect than either *asterie* or *almana*. Of course, these two may very well be distinct forms of one dimorphic insect. This is a very different thing, not in itself improbable ; but Colonel Swinhoe's argument from intermediate varieties tells rather against than for such a theory, and I do not know of any other reason for entertaining it.

20. *J. hierta.*—This is not uncommon in Bombay on the uncultivated parts of Cumballa Hill and about dry stubble fields. It and the next appear later in the year than the preceding species.

21. *J. orithyia.*—This is *par excellence* the *Junonia* of the Deccan, delighting in dry hills and stony plains. On the bare plateau of Lanowlie I have found it very abundant in company with the last, in February, revelling in the wealth of minute wild flowers which clothe the ground in that favoured spot.

22. *Precis iphita.*—After the rains this butterfly is very plentiful, especially among the thorny jungle which covers the little hills of the Konkan. It is also one of the most familiar species on the ghâts. The depth of colour on the underside varies much, and the white spot is sometimes present and sometimes absent. I have never seen specimens here as large as some which come from the Himalayas. It has all the habits of a *Junonia,* and its colour seems inappropriate, for it lives in the midst of green foliage and rarely settles on the ground.

23. *Kallima wardi.*—I believe this grand butterfly is fairly common in every well-wooded part of the country· It appears chiefly in March, April and May, when dead leaves are in fashion, and haunts dry nullahs and ravines, flashing into sight suddenly and as suddenly disappearing into a tree where, after long and cautious peering, you (fail to) discern it sitting motionless on the trunk, inaccessible to your net of course. When you do catch one, it is broken. I suppose their habit of settling in the interior of a tree, upon the trunk or larger branches, tends to break their wings. Last March, the Rev. A. B. Watson, of Poona, made the discovery that this and several other species which most successfully defy the net, such as *Charaxes athamas,* may be captured wholesale at sugar. He had sugared some trees for moths without success, but passing afterwards by daylight, he found that they had become a rendezvous for half a dozen species of butterflies, of which he took as many as he pleased, the present species, in particular, being so infatuated or so drunk that it allowed itself to be taken with the fingers.

24. *Charaxes imna.*—I became aware of its existence of this striking butterfly only last December, when Mr. J. Davidson and I spent part of two days at Matheran in trying to capture two specimens, or rather, I should say, one specimen, for when we got them we found that only half of each remained. I have found since that the species is by no means uncommon on the ghâts from December till March at least ; but it does not put itself in the way of being converted into specimens. It comes out about 10 o'clock, and, selecting a tree with bright shiny leaves, perches bolt upright in the middle of a particular leaf, just a foot above the highest point you can reach with your net. Whether by accident or design, the position is fenced on all sides with a creeper whose sharp-curved thorns lay hold of everything that passes them and let go nothing. There the proud creature sits, chasing away any other butterfly that approaches, and returning to the same leaf. If you pelt it with stones, it darts off, takes a short circuit and returns to the same leaf. You may pelt it for an hour with the same result. You may easily circumvent it, however, by erecting a platform of stones under its perch, but your aim must be sure and your stroke sudden, for no other butterfly goes off with such rapidity. There

is only one specimen of this in the Society's collection, a male which I caught at Khandalla.

25. *Charaxes athamas.*—This is common en' ugh on the ghâts, chiefly, I think, from December to March. It is very similar in its habits to the last, and almost as difficult to capture. They have a *penchant* for certain places, and there seems to be one permanently resident at the reversing station on the Thull Ghât. In the Society's collection there are one or two old specimens of large size, with the apical spot which is wanting in the smaller form.

26. *Charaxes fabius.*—This is not so common as the last, and I know little about it. It occurs in Bombay sparingly. There are four specimens in the collection from Khandesh and the Tanna district.

27. *Cyrestis thyodamas.*—This was very common at Mahableshwar last cold season, from December till March at least. Whether it is usually so I cannot say. I never before met with it, nor heard of its occurrence in the Presidency. I collected a good many specimens, which are decidedly smaller and, I think, better marked than specimens from the Himalayas. It is a sprightly creature, skimming along with the flight of a *Neptis* or an *Athyma*, settling on the upperside of a leaf, with its wings rigidly expanded, then adroitly transferring itself to the underside of the same leaf. It sees remarkably well, but does not settle very high, and is easily caught. I do not think it ever closes its wings, even when it settles on the ground.

28. *Ergolis ariadne.*—I am not sure I have caught this in Bombay, but it is everywhere on the hills during the cold season. It flies low. Mr. Davidson sent me a number of the larvæ from Dhulia in Khandesh in the month of October, together with those of the next, from which they were almost indistinguishable. The pupæ were quite indistinguishable, at least to my discernment. They fed on *Tragia cannabina*.

29. *Byblia ilithyia.*—The specimens in the Society's collection are from Cutch and Dhulia, but I have met with it in Poona. It flies low.

30. *Neptis varmona.*—This species is common enough in Bombay and Poona after the monsoon, and still more so on the hills as late as March. It frequents gardens and hedges, and has a characteristic flight, steady and straight, with jerky strokes of its wings, between which they remain stiffly expanded.

31. *Neptis ophiana.*—I met with a few specimens of this at Mahableshwar last March. It was new to me, but on the wing is so like *Athyma perius* that it may have easily escaped my notice before.

32. *Athyma perius.*—This is common at Khandalla, Lanowlie and Matheran, but I did not find it last March at Mahableshwar, which is 2,000 feet higher. It does not occur on the plains. I found the larva

at Matheran in March, feeding on *Glochidion lanceolatum,* one of the
commonest trees on the hill. This species seems to lay its eggs by
preference on the young shoots that come up from the roots, like *A.
phalanta.*

33. *Euthalia garuda.*—I think this butterfly is less common in the
jungle than it is about human dwellings. It loves to bask on old
grey walls and may be found making itself happy in the dirtiest parts of
the native town. I am quite sure it prefers the liquids which it sips from
the roadside gutter to the nectar of any flower. The larva may be
found in the month of October, and no doubt later, on the mango tree.
I found one once on a rose bush, to which it had done some mischief. It
is a difficult larva to rear, sulking and refusing to feed. It eats only
at night, remaining motionless all day, and the interlacings of its light
green spines form such a perfect imitation of the venation of a leaf that
it must very easily escape detection.

34. *E. lubentina.*—This is not very rare on the hills, but seems
to keep to the tops of trees, basking in the sun. I have found it at
Matheran in December.

35. *Symphœdra nais.*—In structure this is said to be near to *Euthalia;* in habits it is a *Junonia,* or perhaps, I should rather say, a *Pyrameis,*
flying low and alighting generally on the ground, where it basks with
wings expanded. I have not met with it in Bombay, but it is not uncommon at Uran, only five miles from Bombay, and may be found, I dare say,
throughout the low jungles of the Tanna district. I think its chief season
is the close of the monsoon, but I have found it in May. There is some
difference in the depth of colour in different specimens from the same region.

36. *Hypolimnas misippus.*—This is very common after the rains. The
larva feeds on *Portulacca oleracea,* which is a monsoon weed in Bombay.
I think Boisduval's enthusiasm carries him too far when he says that the
mimicry of *D. chrysippus* by this species extends even to their larvæ, which
at first sight have a superficial resemblance. This is a spiny larva of the
Junonia type, and does not need to mimic anything, because nothing is
under temptation to eat it. Females of the *dorippus* type are not rare, and
there is one in the Society's collection.

37. *Hypolimnas avia.*—When the first showers of the monsoon have
fallen in June, a large number of the females of this butterfly appear,
without a single male being visible. Two or three months later, males
appear in great abundance in some places, followed after an interval by
females. I noticed this particularly in 1878 at Uran, where the low
jungle on the hill sides literally swarmed with this species, and I have at
other times, without noting dates so precisely, found one sex abundant
without the other. In Bombay and Poona this species is common about

the close of the rainy season, though never so plentiful as the next. It wanders little, and I have watched a fine male in the garden day after day, basking on the same bush and sucking the same flowers, fiercely chasing all rivals away, until it was old and faded and broken, and finally disappeared. I do not think they live much over a week, but this is a difficult point to settle, because in captivity there are unnatural conditions which may lengthen as well as shorten an insect's life.

38. *H. bolina.*—This is the least common of the three species in Bombay. In collect'ons from Malabar, it is, I think, the commonest. Perhaps it is more a denizen of the jungle and rarer in gardens. Like the others, it appears during the latter half of the monsoon and for a short time after.

A NEW SPECIES OF *ALGA*

CONFERVA THERMALIS BIRDWOODII.

(With an Illustration.)

DISCOVERED AMONG THE HOT-WATER ALGÆ FROM VAJRABAI EXHIBITED BEFORE THE BOTANICAL SECTION ON 15TH MARCH 1886.

By Surgeon K. R. Kirtikar, I.M.D.,

2nd Surgeon, J. J. Hospital,

Acting Professor of Anatomy, Grant Medical College.

I visited the hot-water springs of Vajrâbâi near Bhiwandi in the Thana Collectorate a fortnight ago. The place has been described in the Indian Antiquary of March 1875 (page 66) by Mr. Sinclair, of the Bombay Civil Service, one of our able co-adjutors and generous contributors in the Zoological Section. The springs occur, he says, in or near the bed of the Tansa River at the village of Wadouli, about twelve miles due north of Bhiwandi. Those at Akloli and Ganeshpuri have a temperature of about 100° F. The water is stored, as it bubbles up from the underground springs, in a couple of big basins built of black basaltic stones, about eight feet by twelve in dimensions and four feet deep. The water bubbles up hot through circular holes cut out at the bottom of the basin. It has a sulphurous taste and smell. It was analysed by Drs. Giraud and Haines in January 1855, but no note seems to have been made of this quality of the water. The analysis is given in the Transactions of the Medical and Physical Society of

Bombay (page 24, Vol. V.) and is as follows:—In 10,000 parts or grain-measures :—

Specific gravity at 60° F.	1002·0
Chloride of sodium	12·41
Chloride of calcium	7·07
Sulphate of lime	2.08
Silica	·88
Total Solids...	22·64

The temperature of the water at source is noted 120° F.

The Algæ that I have collected are from the hottest springs of Gorakha-Machhindra, the temperature of which is 130° F.

Very few Algæ are described as the inhabitants of hot springs. Cooke, in his recent work on British Water Algæ, mentions only four— *Stigeoclonium thermale, Gleocapsa arenaria, Spirulina oscillaroides* (variety *Minutissima*), and *Oscillaria thermalis.* Hassall mentions *Oscillatoria thermulis* (page 250, Vol. I., British Freshwater Algæ) as being found in a stream of hot water at Stevenston, but, as Cooke rightly remarks, Hassall's illustration is not sufficiently graphic as to enable the reader to recognize the species. Hassall, however, observes that some of the Oscillatoreæ are found in mineral waters and in such as are absolutely hot and almost boiling.

Kützing, in his work " Species Algarum," describes *Spirulina subtilissima* as being found in some Italian hot springs. The *Spirulina thermalis* is found in the hot springs of Italy and Bohemia (Carlsbad). He also describes, among the doubtful species which he has not fully recognized, *Anabaena thermalis,* found in the Algerian River Ouedel-Hammam, which derives its waters from a hot spring. *Rhizoclonium Crispum* is also described by the same Algologist as being found in the hot springs of Germany and Italy.

Thus it will be seen that the Algal inhabitants of thermal springs are few and far between. I was struck during my visit to Vajrâbâi with the rank growth of the Algæ now exhibited before this Meeting. They were growing luxuriantly, and looked in their recent and natural condition, richly and beautifully green, firmly fixed on to the loose pebbles that were rolling in the stream and to the black basaltic stones lying along the current of the continuously streaming water, the high temperature of which the human hand could not stand for more than two consecutive minutes.

KIRTIKAR's
CONFERVA THERMALIS BIRDWOODII

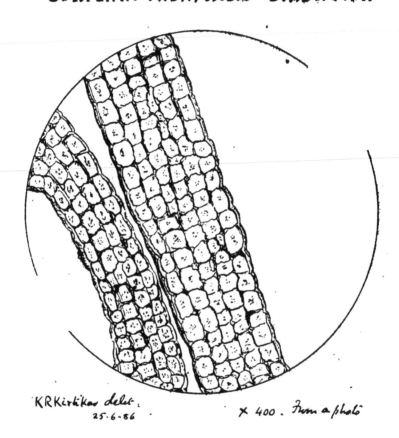

KR Kirtikar delet.
25·6·86

X 400. From a photo

Bombay Natural History Society's Journal —

I found four varieties of Algæ in the different springs about the place :—

(1) A species of Ulothrix, not very different in structure from *Ulothrix Radicans* of Cooke.

(2) A species of Nostoc with its very minutely beaded appearance.

(3) A Conferva very similar to *Chœtomorpha implexa* (*vide* p. 140, Cooke's Algæ, plate 54, fig. 6).

Those three varieties will by and bye receive special treatment at my hands, but to-day I propose to examine in detail the fourth species of Conferva which I have not seen described anywhere in Kützing, Cooke, Hassall or Mrs. Gatty.

(4) To the naked eye this variety of Conferva is visible in the shape of fine hairy filaments of beautiful rich green. Under the microscope with a $\frac{1}{4}$ inch objective the structure is seen in detail, and is not unlike that of *Enteromorpha Percursa* described and figured by Mrs. Alfred Gatty in her British Seaweeds under No. 350, Plate LXXII, the difference being that our specimen has distinct dissepiments in the body of the Alga and tretrasporous arrangement of the zoospores inside the tubular segments. Following Mrs. Gatty's mode of description given in her work, I here briefly give the result of my examination of the newly-discovered Alga.

Color.—Bright rich green when in the hot water ; turning olive green on being kept in cold water, or on drying.

Substance.—Soft ; can be easily torn off.

Character of frond.—Single ; bearing occasionally slender spine-like branchlets, short and tapering, not distinctly jointed, growing in tufts.

Joints.—Small, numerous, faintly marked, with from four to six dissepiments in the long axis of the Alga. Smaller horizontal joints separating the endochrome and zoospores into spaces $\frac{1}{1,000}$ to $\frac{1}{3,000}$ inch in length and $\frac{1}{1,000}$ inch in breadth.

Measurements.—4 to 6 inches high when standing in the hot stream in tufts.

Fructification.—Unknown. In some of the mature segments the central mass of coloring matter constituting the sporidium is arranged in a tetrasporous manner, the contents escaping in due time, probably by rupture of the segment walls.

As the species is quite a new one, requiring a designation, I have obtained the Honorable Mr. Justice Birdwood's permission to associate his name with the Alga, as he is at present the President of our Botany Section; and as I wish to mark the high sense of esteem and respect I entertain for him as an accomplished and practical naturalist, and as a kind and indefatigable worker in the interests of humanity, I call the Alga *Conferva Therm̐lis Birdwoodii,* and so be it known in the weedy world.

K. R. KIRTIKAR.

NOTE ON FREQUENCY OF PARASITES IN INDIAN ARMY HORSES.

By V. S. J. H. Steel, A.V.D.,

Supdt., Bombay Veterinary College and Hospital.

In August 1884 I examined with care the bodies of twelve Light Cavalry horses destroyed on account of age or incurable injury at Bangalore. The results are, in some respects, remarkable. Thus, no doubt is left as to the richness of the zoological field explored by me; every one of these twelve horses contained large numbers of parasites of two or more species. Further, a young mare, the only Australian of the lot, had two forms of parasite which were not fouِid in any of the others; this suggests the question as to whether she can have brought those forms from the depôt at Oossoor some two years before, or from Australia some three years before, her destruction. Again, certain parasites commonly seen during *post-mortems* of horses were conspicuous by their absence; echinococcus cysts were not found in the liver, nor armed strongyles in the anterior mesenteric artery, nor were any thread-worms present in the respiratory passages, nor flukes in the liver. If well-fed and cared-for horses were thus infested, how much more so must be country ponies and horses "roughing it" out in the districts. All the horses had been watered from the same tank; had been standing in open lines; and had been similarly fed for the six months previous to destruction. Their fodder mainly consisted of fresh Hariali or Dhoob grass, more or less moist from washing, but fairly well cleaned as regards removal of mud, dirt, and foreign grasses. Their gram was boiled coolthee.

Parasites found in various situations.

No.	In Stomach.	In Cæcum.	In Colon.
1	Bots and small round-worms.	[One ascaris megaloce-phala in duodenum]	Oxyurides(and in rectum).
2	Bots................:.........	Str. armatus and str. tetracanthus.	
3	A few bots	Str. armatus	Str. armatus and amph's-toma collinsii.
4	A very few bots	One worm cyst	Amphi. collinsii (at com-mencement).
5	Bots, a large cyst, a few small thread-worms.	....	Str. armatus (a few), str. tetracanthus, and amphi. collins i (many) at com-mencement of double colon.
6	A number of bots and of small thread-worms.	Str. armatus and str. tetracanthus (imma-ture ?) ; also amphis-toma collinsii.	Str. armatus and str. tetracanthus (imma-ture?) ; also amphi. col-linsii.
7	Bots	Str. armatus, str. tetra-canthus (a few imma-ture ?) ; also amphi. collinsii (some).	Str. armatus, str. tetra-canthus (a few imma-ture ?), amphi. collinsii (some)at commencement of double colon.
8	Many bots, a few small thread-worms.	A few mature str. armatus.	A few mature str. armatus.
9	A few bots...............	A few str. armatus	Str. armatus and amphi. collinsii (a few) at com-mencement.
10	Many bots	One small cyst, str. tetra-canthus ? and a few amphistomes.	Amphistomes, str. arma-tus, and str. tetracanthus (in enormous numbers).
11	Bots and an enormous worm tumour.	A few tumours and small thread-worms (str. tetra-canthus ?).	Numerous amphistomes.
12	Bots and many small round-worms.	One tænia, many str. armatus.	A few amphistomes ; blood spots as though from parasites.

The evidence given here is of two kinds : (*a*) Positive—including presence of parasites or indications of their having been present ; and (*b*) Negative—the absence of parasites and of traces of them.

We have positive evidence of the occurrence of parasites as follows:—

1. Bots, *the larvæ of œstrus equi,* in the stomach only, from which we infer that these partial parasites are at Bangalore in August not yet ready to assume the chrysalis stage. We can easily understand why bots were present in every case examined, for all the horses were standing in open lines and fastened by head-and heel-ropes ; they were, therefore, continuously exposed to the attack of the gad-fly and deprived of power to escape it by flight into water or otherwise. The horse gad-fly is, however, not very irritating, and it deposits its eggs on the long hairs of the legs instead of on more sensitive parts, such as the nostrils, attacked by the gad-fly of the sheep, and (by puncture) the skin as in case of the ox gad-fly.

2. *Small stomach thread-worms*, present with or without cystic " abodes." These are representatives of large-mouthed or small-mouthed *spiroptera* (or of both). It is the large-mouthed form which occurs in the cysts. Of these latter, one was closed and two were open. In four cases these small thread-worms were found, but no trace of cyst ; these were probably the small-mouthed form ; unfortunately no microscopical examination was made to settle this point.

3. *Ascaris megalocephala,* in only one case out of twelve, is probably considerably below its frequency among horses in England. The specimen was small and apparently immature.

4. *Oxyuris curvula* in only one case. This parasite, the presence of which is denoted by a white or yellowish deposit of ova around the anus of the host, is of frequent occurrence in the rectum of country ponies, and certainly is not rare in India. Its infrequency in these horses was probably due to this not being its " season" for abode in the rectum, or, as this host was a " Waler" and young as compared with the most of the other horses, the oxyurides may have been brought from the Oossoor Depôt or from Australia. In support of this latter view is the fact that the parasite is very frequent in recently imported " Walers," but opposed to it is the fact that the host had been some two years in the ranks and nearly three years in this country.

5. The single tape-worm observed was apparently a *Tænia perfoliata ;* it was a wretched specimen obtained from a Persian horse which had been some eighteen months in the country and had been marched from Bombay to Bangalore after purchase. I am inclined to think he brought this tape-worm with him, perhaps from Persia. This species of parasite is frequent among asses in England and not rare in the horse. The blood spots on the lining membrane of the colon in the case from which it was taken may have resulted from previous and recent occupation by other individuals of this species, but there was no sign of more than the one which was met with in the cæcum. Amphistomes were also present in this case, but they seldom cause blood spots. The effects of involuntary change of country by parasites on import or export of their hosts would be an interesting study—which of them in their unintentionally adopted countries find the complex requisites for their strange metamorphoses in development remains to be established.

6. *Strongylus armatus* was in five cases found in both cæcum and colon, twice in the cæcum and not in the colon ; twice in the

colon and not in the cæcum. In all cases the parasite was in the mature form, and in no case was the larval armed-strongyle found in the anterior mesenteric artery causing "worm aneurism." It is evident that in August in Bangalore the strongyle is in the adult dung-eating stage and found in the large intestine—whether this is the case in the rest of India remains to be proved. These are the worms considered by Böllinger as a frequent cause of colic.

7. The evidence about *Strongylus tetracanthus*, though conclusive as to presence in some cases, is not invariably satisfactory. The parasites seen were small thread-worms of a white color (entered as " immature *str. tetracanthus* " in my rough records as prepared at the time). They seldom (*i.e.*, in only one case infesting the cæcum and in one infesting the colon) had the distinctive red colour of *str. tetracanthus*. They more resembled spiroptera in four cases of the cæcum and three of the colon. Not in a single case did I find the form which has been called *trichonema arcuata*, *i.e.*, the young *str. tetracanthus* forming small rings in the substance of the mucous membrane of the cæcum and commencing portion of the colon. But in one case was a cyst of the cæcal mucous membrane, and in two other instances where the small white worms were *not* found were cysts, one burst, the other unopened. No microscopical diagnosis, unfortunately, was made of the " small white worms," so *we must leave it an open question whether a form of spiroptera is found in the cæcum of the horse.* The cysts may possibly have resulted from migrating *str. armati*.

8. *Amphistoma collinsii*, a form of trematode, was found in the cæcum in three cases, but in the colon in no less than nine. From this we might infer that the latter is specially its habitat. The commencement portion of the colon is most frequently invaded. I would in this connection suggest the view that *frequency of parasites such as are introduced with the food or water is found in the former case in the stomach, in the latter in the cæcum or commencement of the duodenum.* This is a generalization of considerable importance and worthy of discussion ; if it be accepted, we may infer that the amphistomes in their larval form are ingested from the muddy water of tanks either free or in the textures of minute larvæ. These amphistomes are very common in horses throughout India.

It is remarkable how much freer from parasites some horses are than others. No. 10's intestines and stomach constituted quite a zoological garden for the helminths. It is remarkable that, even leaving bots out of the question, in not one of these twelve cases was there freedom from parasites.

I beg to be permitted to close this paper by recommending the interior organs of domesticated and other animals to members of the Society as a "happy hunting ground." There is enormous scope for research, and material in every kitchen and every butcher's establishment. The odour of gastric and intestinal contents may not be so enticing as that of the hill air, the ocean breeze, or the fresh dry atmosphere of the maidan, but the aroma and gases from animals' bowels are harmless, and (I speak from experience) make one wondrous hungry! Again, consider the importance of the problems to be solved ; *in every part of the world the same endo-parasitic species are subjected to the same surrounding conditions of food, temperature, and reaction ; any specialities of geographical range must depend on influences from without ; thus our search for causes of parasitic invasion should be limited in its range and much facilitated.* The remarkable biological phenomena observed in study of the life-history of parasites, and their considerable influence on the health and even life of those higher animals they occupy as "guests," render them a specially interesting study to the medical or veterinary worker.

J. H. S.

LIST OF BIRDS COLLECTED AND PRESENTED TO THE SOCIETY
By Mr. A. T. H. NEWNHAM, S.C., 10th N.I.

1.	Neophron ginginianus	White Scavenger Vulture.
2.	Falco peregrinus	Peregrine Falcon.
3.	Astur badius	Shikra.
4.	Accipiter nisus	Sparrow Hawk.
5.	Aquila vindhiana (eggs)	Tawny Eagle.
6.	Hieraetus pennatus	Booted Eagle.
7.	Circaëtus gallicus	Bonelli's Eagle.
8.	Nisaetus fasciatus	Crestless Hawk-Eagle.
11.	Butastur teesa	White-eyed Buzzard.
12 to 15.	Circus macrurus	Pale Harrier.
16, 17.	Carine brama	Spotted Owlet.
18.	Coracias indica	Indian Roller.
19 to 21.	Halcyon smyrnensis	White breasted Kingfishes.
22, 23.	Ceryle rudis	Pied Kingfisher.
24.	Palæornis torquatus	Rose-ringed Paroquet.
25.	Yunx torquilla	Wryneck.
26.	Eudynamis honorata	Indian Koel.
27.	Cinnyris asiatica	Purple Honeysucker.

28. Lanius lahtora ....Grey Shrike.
29. Lanius erythronotus ....Rufous-backed Shrike.
30. Lanius vittatus ....Bay-backed Shrike.
31, 32. Lanius isabellinus..............................Pale Shrike.
33, 34. Pericrocotus erythropygius ....White-bellied Minivet.
35. Pyctoris sinensis ....Yellow-eyed Babbler.
36. Thamnobia cambaiensis ....Indian Robin.
37. Pratincola caprata ....White-winged Bush-chat.
38, 39. Pratincola indica ....Indian Bush-chat.
44, 45. Cyanecula succica ....Blue Throat.
46. Franklinia buchanani ....Rufous fronted Warbler.
47. Franklinia buchanani ....   Do.   do.
50. Motacilla Maderaspatensis ....Large-pied Wagtail.
51. M. dukhunensis ....White-faced do.
52. M. leucopsis ....
56, 57. Agrodoma campestris...........................Stone Pipit.
58. Agrodoma sordida ....Brown Rock Pipit.
59, 60. Gymnoris flavicollis ....Yellow-throated Sparrow.
61. Emberiza striolata ....Striolated Bunting.
62. Mirafra erythroptera.................................Red-winged Bush Lark.
64. Alaudula raytal ....Indian Sky Lark.
65. Spizalauda deva..Crested Lark.
67, 68, 69. Pterocles arenarius ....Large Sand Grouse.
70. P. fasciatus ....Painted Sand Grouse.
71, 72, 73. P. senegallus................................Spotted Sand Grouse.
74, 75, 76. P. exustus ....Common Sand Grouse.
77, 78. Francolinus vulgaris.........................Black Partridge.
79. Perdicula asiatica ....Jungle Bush Quail.
80. Houbara McQueenii ....Houbara.
81. Cursorius gallicus ....Cream-colored Courser.
82. Chettusia gregaria....................................Black-sided Lapwing.
83. Lobipluvia malabarica ....Yellow-wattled Lapwing.
84. Œdicnemus scolopax................................Stone Plover.
86. Totanus ochropus ....Green Sandpiper.
87. Totanus glottis ....Green Shank.
89. Himantopus candidus ....Stilt.
90. Fulica atra ....Coot.
91, 92. Ardetta sinensis ....Yellow Bittern.
93. Botaurus stellaris ....Bittern.
94. Dendrocygna javanica ....Whistling Teal.
95. Anas boschas ....English Mallard.
96, 97. Chaulelasmus streperus ....Gadwall.
98. Mareca penelope ....Widgeon.
99, 100. Querquedula crecca ....Teal.

Additional.

103. Elanus cæruleus.......................................The Black-winged Kite.
104. Rhynchæa bengalensis. ....The Painted Snipe.
105. Gallinago gallinaria ....The Common Snipe.

ZOOLOGICAL NOTES.

NOTE ON AN *OLIGODON* (*SUBPUNCTATUS*?) FOUND AT DAHANU, NORTH KONKAN, MARCH 1886.

BY MR. G. VIDAL, C.S.

Description.—Length 11¼. Scales 17. Upper labials 8 (4, 5, 6 entering orbit). Minute black spots on the dorsal line about every third scale *not* white edged.

A lateral streak of minute black specks.

Scuts black spotted on each side.

This specimen agrees with the description of *Subpunctatus* (D *et* B), except that the scales are in 17 rows and not 15, and that the dorsal spots are plain and not white edged.

In the number of rows of scales it agrees with *Spini punctatus* (Tan), but the latter, according to the description, has 9 upper labials and no ventral dots.

G. VIDAL.

PTEROPUS EDWARDSII.—One of the 21st of May, one of the hottest days, I suppose, that man has endured on this side of India, I was at Belapur near Panvel, and at about 1 o'clock in the day I came upon several trees covered with Flying Foxes, all wideawake and *fanning themselves* hard with one wing. Some used the right wing and some the left, but not one was at rest. More than a hundred wings waving at once produced a very striking effect, and I cannot think that the habit, if at all general, can have altogether escaped notice. I am curious to know if anyone else has observed it.

E. H. AITKEN.

WHITE-ANTS.—The following seems worth noting. I have heard of similar cases, but this is the first that has come under my own observation. One of the windows of the travellers' bungalow at Panvel had been attacked by white-ants, when it was opened and left open for two days, thus cutting them off from their base of operations. Instead of working along to the side of the window and going down by the frame, they had made an earthen pipe, three inches long, to connect the window with the sill below. The pipe was perfectly straight, like a mill chimney, and very thin, just wide enough to allow passage for one ant at a time ; so they must have had some arrangement for obtaining "line clear" before entering at either end. White-ants being blind, it is an interesting question by what sense they assured themselves when they commenced their pipe that they were not working out into space.

E. H. AITKEN.

Editor's Note.—A chest of drawers was removed about 4 or 5 inches away from a wall. The feet of the chest were inserted in saucers of turmeric powder, and the contents were considered safe. But on opening one drawer after a time, it was found full of white-ants. On looking behind the chest, there was discovered a track leading up the wall to a level with the drawer, and then a bridge consisting of a single pipe was thrown across and the drawer entered.

R. A S.

POISONOUS LIZARDS, THE BIS-COBRA.

Editor's Note.—In a letter to Mr. Phipson, Honorary Secretary to the Society, Mr. Ommanney, Under-Secretary to the Government of Bombay, states that in the official reports seven deaths in Guzerat are put down as having been caused by a poisonous lizard. He supposes this to be the much-discussed Bis-Cobra, and asks for information concerning it or any other poisoncus lizard, if such a thing exists in this part of the world. Mr. Phipson replied that " all naturalists are of opinion that no such thing as a poisonous lizard exists in this country. The belief to the contrary is, however, prevalent in India amongst the ignorant classes in country districts, and is doubtless kept up by the snake-charmers and others whose interest it is to foster public credulity in such matters.

"The word Bis-Cobra is applied to a variety of lizards in different parts of the country, but in all cases where the reptiles have been pointed out by the natives and killed, and sent to museums, they have been at once identified as known species." He adds in a postscript " that according to the highest authority the only lizard the bite of which is known to be poisonous is the Heloderma of the S. W. States of America and Mexico." It is doubtful now whether the venom of the Heloderma is as powerful as has been reported. I believe no authentic case has been known of the death of a human being from its bite, though small animals suffer to a fatal extent. I have never seen any lizard in India like it; any sort of lizard may be a Bis-Cobra to a native. I once saw a whole *Kacheri* full of people put to flight by a common garden monitor. From what I remember of the Heloderma which was presented to the London Gardens by, I think, Sir John Lubbock, the nearest approach in form is our *Uromastix hardwickii*, only flatter, and yellow and black instead of earthy brown, the whole body covered with small tubercles ; a very repulsive looking creature, and capable of giving a severe bite. I believe it killed some small animals : Guinea-pigs and the like. It arrived in a tin box long and narrow, and when this was opened at the end it would not come out, but planted its claws against a ledge at the opening and refused to budge. I think it was Mr. Bartlett himself who told me that, not believing in its poisonous properties, he caught it by the head and pulled it out.

R. A. S.

ON CONJUGAL INFIDELITY AMONG BIRDS.

By Mr. W. E. Hart.

I was interested some weeks ago in reading in the pages of "Nature" several accounts of instances of conjugal infidelity among birds. Curiously enough a somewhat peculiar case came under my own observation shortly afterwards. About the end of April a pair of wild pigeons, in appearance resembling the " blue rock " of England, began to build their nest in my porch on the top of one of the pillars supporting it. One night, before the nest had been completed, the hen bird was attacked in her sleep by some beast (I suppose a rat) which bit off one of her legs. She did not seem much worse for the loss, but from that time nothing seemed to go right with the nest. It was constantly falling to the ground. On two occasions after an egg had been laid in it. At first I thought this was due to the crows, but I think now it may have been caused by the awkwardness of the hen bird, in her mutilated condition, when alighting on or rising from her nest. In vain the unhappy pair time after time repaired the disaster, shifting the position of the nest from one corner to another till they had tried all four pillars. When we went to Matheran in May the nest was still unfinished, the eggs still unlaid, and there seemed no chance of our unfortunate friends ever succeeding in raising a brood of chicks. Still we could not but admire and sympathize with their patient, persevering industry and fidelity to each other in adversity, and recal the traditions we had heard of how the pigeon, the emblem of love, mates for life, and how, when death takes one of the fond couple, the survivor pines away and dies of grief. Alas! for another shattered illusion! When we came back from Matheran we found the nest finished indeed, and tenanted by a pair of well-grown chicks nearly fledged, but they were not the children of our one-legged friend. Her faithless spouse had brought home a second bride with the proper complement of limbs, who now ruled his house, accepted his caresses, and regulated the affairs of his nursery, while the first looked sadly on, standing sorrowful and solitary on her one leg. She, poor thing, apparently cannot get it out of her head that she is the true wife and real mistress of the house, for she often tries to approach the nest or the chicks. But as often as she does so, her rival flies at her and drives her off, and even carries her hostility so far as to attack her unprovoked when she is sitting quietly by herself at a distance. Lothario, I am glad to say, never joins in actively ill-treating the deserted one. But his coldness and neglect must be as hard to bear. As she never leaves the neighbourhood, I can only hope her forlorn appearance acts as a perpetual blister to his conscience.

W. E. H.

BOTANICAL NOTES.

NOTE ON THE *FERONIA ELEPHANTUM* (ELEPHANT OR WOOD APPLE) AS A TIMBER TREE.

By MR. FRANK ROSE.

N. O. RUTACEÆ (Aurantiaceæ, or Orange Order.)

This apparently insignificant Indian tree seems not to have found a description in Balfour's " Class Book of Botany, 1854 ; " yet a Botanist in 1829 deemed it a " noble Indian tree." Be that as it may, besides being a medicinal agent, its properties, I think, are so well known as to need no reiteration in this journal; suffice it to say that every part of this " common jungle tree " is reputed to be useful. It was gracefully named " *Feronia* " after the " Goddess of Forests " by the celebrated Portuguese Botanist CORREA DE SERRA. My object in writing on this subject is to question the assertion of a respected writer, who states that the timber is "*used for house building;* " probably he meant cnly for *temporary* structures? I write from experiense, and beg to differ from him. A beam of this wood, to save expense, was put up in a bungalow in 1880, and in 1886 perforations by borers were the result! I anticipated this, and informed the builder at the time that a certain percentage of saccharine matter is contained in this tree, consequently, it was open to the ravages of insects. The timber is certainly tough, the average weight per c. ft. = 49 lbs.; is almost equal to that of teak (*Tectona grandis*) ; it planes smoothly and receives a good polish ; but *cui bona?*

En passant, CREOSOTE, possessing that powerful antiseptic property, has been recommended for the *preservation of timber ;* but instances have occurred where creosote, chloride of zinc, carbolic acid and corrosive sublimate have been used, but without satisfactory results, excepting that they retard the destruction by insects for a couple of years or so, when the above have not penetrated the wood. Creosote, I know, acts like a charm, and is efficacious in *preserving animal substances.* Then in my humble opinion I consider that the wood of the *Feronia* is unfit for permanent structures, though it may be used for agricultural implements, but should not be classed with the " Indian timber trees " of durability.

F. R.

Note by Editor.—The *Feronia, Koit,* or *Katth bel* is mentioned in Balfour's ‹' Timber Trees of India," 1862 Edition, and he reports it as much used for building in Gujerat and Coimbatore, where it is said to be durable, but in Vizagapatam, where it is also much used, it is said to be not very durable, thus confirming Mr. Rose's opinion. Its strength (360 lbs.) is apparently almost equal to teak, but there the comparison ceases ; the durability of teak, its properties of resisting insects, and preserving iron from rust are chiefly due to the amount of tar contained in the wood ; this tar, which was first brought to my notice by the Gipsies (*Bunjaras*) of the Central Provinces in 1863 or 1864 was sent by me for analysis to the Agri-Horticultural Society of Bengal, and the report will be found in the journal of that time.

R. A. S.

PROCEEDINGS OF THE SOCIETY DURING THE QUARTER.

THE usual Monthly Meeting of the Society was held on Monday, April 5th, at 6, Apollo Street, Dr. D. Macdonald, Vice-President, presiding.

The following new Members were elected :—H. H. the Maharaja Saheb of Indore, Sir Jamsetjee Jeejeebhoy, Bart., Captain Street, Mr. N. R. Cumberleye, Captain L. L. Fenton, Captain W. Aves, Mr. G. W. Terry, Mr. J. Franklin, Captain Barclay, Captain Bishop, Mr. H. Van Bnith, and Mr. D. George.

Mr. H. M. Phipson then acknowledged receipt of the following contributions to the Society's collections during the past month, and made a few explanatory remarks regarding the specimens :—

Contribution.	Description.	Contributor.
A number of snakes (from B. Burma)	*Bungarus fasciatus* ...... *Naga tripudians* .... (*Var Maccra Keantiah.*) *Cylindrophis rufus* ...... *Xenopeltis unicolor* ...... *Simotes tæniatus*	Capt. C. H. Bingham. Do. Do. Do. Do.
2 Flying squirrels (alive)	*Pteromys oral* .... (*Var. Cineraceus.*)	Genl. J. Watson, V.C., Baroda.
A quantity of insects from Ceylon.	....	C. A. Stuart.
Skin of a Flying Squirrel	*Pteromys oral*	Do.
A quantity of fresh-water fish from Savitri River.	....	W. Sinclair, C.S.
1 Snake.............................	*Oligodon subpunctatus.*.	Do.
1 Turtle	*Sp. Emys*	Do.
A quantity of sea anemones from Dharamtar.	....	Do.
2 Sea eels	*Sp. Muræna*...............	W. H. McCann.
Fossil tooth of elephant from Rangoon.	....	Wm. Shipp, C.E.
3 Oryx heads from Africa......	....	Capt W. Aves.
5 African gazelles' heads from Africa.	....	Do.
1 Spring bok head from Africa.	....	Do.
1 Fresh-water turtle...............	*Sp. Emys*	G. W. Vidal, C.S.
1 Dotted skink	*Eumeces punctatus*	Do.
1 Snake	*Oligodon subpunctatus.*.	Do.
2 Kangaroo rat (alive)............	*Hypsiprymnus rufescens*	G. F. Johnson.
A quantity of corallines	....	Mrs. Hart.
1 Fresh-water fish.................	*Wallago Attu*	H. H. Swan, C.E.
A quantity of fish & scorpions from Suakim.	....	H. Wenden, C.E.
Megapodius Nicobariensis and egg from Nicobar Islands.	....	Do.
1 Alligator's skin	*Crocodilus porosus*	Mrs. Sleater.
A quantity of algæ	....	Capt. Bishop.
Corallines and marine specimens from Persian Gulf.	....	Do.
14 Fossil Echinidæ from Khavey Island.	....	Do.
A quantity of Iron Pyrites from Larek Island.	....	Do.
Specimen of sponge and coral from Suakim	....	Col. Walcott.
Fossil tooth of elephant from Runn of Cutch.	....	H. A. Acworth, C.S.
Sucker Fish	*Sp. Echeneis*	Do.
1 large Dhaman (alive)	*Ptyas mucosus*............	Wm. Shipp, C.E.
Skin of white-bellied flying squirrel	*Pteromys albiventer* ...	R. A. Sterndale.
Skin of yellow cheeked Marten	*Martes Flavigula*	Do.

Minor contributions from H. A. Acworth, C.S, Mrs. H. S. Symons, J. Parmenides, H. Buckland, Chas. Lowell, C E. Crawley, W. W. Squire, Mrs. A. F.

Turner, Father Dreckmann, W. Gleadow, William Shipp, R. Wroughton, and Dr. E. M. Walton.

Contributions to the Library.—"Vegetable Materia Medica of Western India" (Dr. Dymock) from author; "Game Birds of India, Burmah and Ceylon" (Hume Marshall); W. Sinclair, C.S.; "Moses and Geology" (Kinns) W. Sinclair, C.S.; "Wanderings of Plants and Animals" (Hehn), W. Sinclair C.S.

Specimens deposited with the Society.—5 Cashmere stag heads, from Dr. Banks; 2 Himalayan Ibex heads, from Dr. Banks; 1 Ovis Ammon head, from Dr. Banks; 1 Markhor's head, from Dr. Banks; 1 Cheetal's head, from H. S. Wise; 2 Black Buck's heads, from H. S. Wise; 1 Hyena's head, 1 Wild Cat's head, 2 Neilghai heads, mounted by the Society's taxidermist for up-country correspondents, were also exhibited by Mr. E. L. Barton.

The Secretary announced that the second number of the journal, containing much interesting matter, was now ready for issue to subscribers. A vote of thanks was passed to Mr. Sterndale for having, in the absence of Mr. Aitken, undertaken the sole task of editorship and for bringing out the journal so punctually.

Mr. Sterndale exhibited, through the courtesy of Messrs. William Watson & Co., two cubs of the Indian Sloth Bear (*Ursus labiatus*,) the property of Mr. Mainwaring, and now on their way to the London Zoological Gardens. One of these cubs is the Albino referred to in the second number of the Society's Journal. The cubs were taken out singly and petted by some of the members present, who were much amused at the petulant cry, like that of an infant, which the little bears made when separated.

The living Flying Squirrel presented by General J. Watson, V.C, was also exhibited, and appeared to be none the worse for its flight across Rampart Row, recently alluded to and described in the Journal.

Mr. Sterndale also exhibited a very tame specimen of the Mongoose Lemur (*Lemur mongos*) from Madagascar.

Mr. Phipson turned loose one of the two Kangaroo rats (*Hypsiprymnus rufescens*) lately received from Mr. G. F. Johnson, of the P. and O. Company, Adelaide. The little animal, which is about as big as a rabbit, went bounding round the rooms and caused much amusement to those present.

Dr. Maconachie showed, under the microscope, a sample of the Tulsi drinking-water, collected an hour or two before the meeting. Among masses of vegetable matter there were crustaceans, worms, infusoria, animalcula, and other animal specimens, living, dead, and in various stages of decay.

THE usual Monthly Meeting of the Society was held on Monday, May 3, at 6 Apollo Street, and was largely attended. The Hon Mr. Justice Birdwood, Vice-President, took the chair. The following new Members were elected: His Highness the Rao of Cutch, Colonel F. G. Wise, Mr. C. L. Weber, Mr. R. N. Mant, Mr. G. P. Millett, Mr. James Tod, Mr. R. Riddell, R.E., Mr. Robert Clark, Mr. James Cheetham, Mr. T. R. Booth, Dr. D. G. Dalgado, Rev. H. P. Le Febvre, Mrs. Charles Douglas, Mr. P. C. Oswald, Rev. J. Forgan, Mr. Thomas Bromley, Mrs.

Dillon, Mr. T. W. Pearson, Mr. R. H. Macaulay, Mr. Andrew Hay, Dr Bhiccajee Eduljee Gaswalla, Mr. C. E. Fox, Mr. Montagu C. Turner, and Mr. A. E. Hoare.

Mr. H. M. Phipson then acknowledged the following contributions to the Society's collections during the past month :—

Contribution.	Description.	Contributor.
A Golden Pheasant alive from China	......	Mr. E. D. Barton.
A number of Snakes and Lizards from Saugor ...	......	Lieut. Barnes.
Snout of Saw-fish............	*Pristis anteguorum*	Miss R Rich.
2 skins of Flying Squirrels	*Pteromys orul*	Mr. E. C. K. Ollivant.
A number of Snakes and Reptiles from Deolali...	......	Mr. F. C. Webb.
Head of large Saw-fish....	......	Mr. D. E. Aitken.
1 Porcupine alive	*Hystrix leucura*	Mr. A. S. Ritchie.
Head of Sind Ibex	*Capra ægagrus*..............	Mr. B. T. Ffinch.
Head of Bison...............	*Gavæus gaurus*	Mr. Robt. Clark.
3 Snakes	*E c h i s Carinata, Dipsas Gokool*	Mr. F. D. Campbell. C.E

Minor contributions from Mr. H. F. Hatch, Dr. Kirtikar, Mr. E. H. Aitken. Mr. J. W. Evans, and Mr. Nanabhoy Rachanath.

Contributions to the Library.—A series of photographs of animal shot, by Mr. J. D. Inverarity. A paper on a new Gerbillus (by Mr. James Murray) from the author. A paper on a new species of Mus (Mr. James Murray) from the author.

Mr. Phipson also announced that he had received a telegram from H. H. the Maharaja Sahib of Indore, offering the Society two panthers, which had, however, not yet arrived.

A vote of thanks was then passed to the ladies and gentlemen who had sent the following exhibits to the meeting :—

Exhibit.	Description.	Exhibitor.
1 Orchid	*Phaleonopsis grandiflora*...	Mr. M. C. Turner.
1 Do.	*Saccolobium guttatum*	Mrs. Chas. Douglas.
1 Do.	*Dendrobium pierardi*	Mr. M. H. Starling.
3 Do.	*Ærides crispum*	Mr. E. H. Aitken.
2 Do.	*Dendrobium nobile*	Victoria Gardens.
1 Do.	*Adiantum peruvianum* ...	Captain Passy.
1 Fern	*Adiantum concinnium*......	Mr. M. H. Starling.
1 Do.	*Adiantum fergusonii*	Do.
1 Do.		Do.
A quantity of new rare plants	......	Mr. T. Bromley.

Mrs. Charles Douglas, Mr. G. W. Terry, and Mr. R. A. Sterndale showed some beautiful drawings of plants. Mr. E. L. Barton exhibited a carpet made from a tiger's skin and twenty-two black buck skins, and also a large specimen of Rock Snakes (*Python molurus*) stuffed and mounted by himself.

Mr. Starling drew attention to a fern (*Adiantum ferguscnii*) exhibited by him, and explained that it had been found about five years ago in a garden at Negombo in Ceylon, but that no one knew how or whence it had come there. The species was unknown at Kew, but the authorities there considered that it was a cross between *A. farleyense* and *A. tenerum*. Looking, however to the place where it was found, that was impossible, as *A. farleyense* had never been known to bear spores in Ceylon. It was, therefore, regarded by Dr. Trimen, the Director of the Botanic Gardens at Peradeniya, as a new species, and named by him after the discoverer, Mr. Ferguson, the Municipal Engineer in Colombo. Mr. Starling also suggested that it would be useful if those who had plants of *A. farleyense* would watch them, as his had apparently prepared to bear spores, the edges of the leaves having turned under, to as to form receptacles, but that he had not hitherto been able to detect any spores.

JOEY.— See Reversion to primitive types.

JOURNAL
OF THE
BOMBAY
Natural History Society.

No. 4.   BOMBAY, OCTOBER 1886.   Vol. I.

WATERS OF WESTERN INDIA.
PART II.—KONKAN AND COAST.
(By a Member of the Society.)

THE region of the present paper is included, roughly speaking, between the 16th and 21st degrees of North Latitude and between the watershed of the Sahyadri Range, with an average elevation of about 3,000 feet (rising in places to 4,500) and the outer line of soundings, where they increase suddenly, though very irregularly, at a distance of about 60 nautical miles from the coast.

The mountains, the coast, and the line of deep water are pretty nearly parallel, running from south-east to north-west, with a slight westerly divergence in the coast-line and a more marked one in that of soundings.

The whole region forms the face of the Deccan trap area, descending westwards into the ocean by a series of the terraces or steps which characterize this formation and have given it its name (trappa=step in Swedish or Danish).

Fresh and salt water are so much mixed up in parts of this region that it is convenient to take the whole together in rough notes like the present.

Between the crest of the Sahyadris and the edge of the series of cliffs which form most of their western face is a narrow highland zone called the " Konkan-Ghát-Mátá," or " Konkan on the top of the gháts." "Mátá" in Maratha means the top of anything, from a skull to a mountain, whence, for instance, Materan ("The jungle on the hill-top").

The longest torrent of the Konkan-Ghát-Mátá is probably the Kumbhe nullah, with a course of five miles ; and I suppose that the little tank at Khandala is its largest sheet of standing water. The torrents, which are very numerous, generally contain water here and there throughout the year : in potholes under falls, or at spots where springs occur in their beds.

These are inhabited by characteristic little fishes ; loaches (*Nemachili*) and mountain carps (*Discognathi*). There do not seem to be many species. I could only distinguish two loaches and one cyprinoid amongst many hundred specimens collected from every spring and stream in the basin of the Savitri. The cyprinoid seems to have the characters of Dr. Day's *Discognathus* (*olim Mayoa*) *modestus*, a species which he bases upon two specimens in the Calcutta Museum, and supposes to belong to Northern India. Lieutenant Beavan remarks on its similarity to his *Dicognathus macrochir*.

One of the loaches is apparently *Nemachilus rupelli*; the other I could not identify. All three seem to live chiefly on green water mosses coating the stones of the streams ; but they are probably pretty omnivorous. They form a sort of Alpine club; there is no tiniest spring that does not hold them ; and the hillmen all maintain that they ascend by leaving the rivers during the rains, and literally climbing up the mountain sides at that time streaming with water.

From some experiments that I made, I think that this extraordinary statement is probably true. The most remarkable other inhabitants of the Ghát-Mátá waters are certain highland periwinkles (*Cremnoconchi*), whose resemblance to the Marine Littorinœ (which people buy by the pint and eat with pins) has given rise to conjecture that they may be descended from "winkles" that inhabited the Ghâts when these were washed by a prehistoric ocean. They seem to sleep in concealment during the dry weather, and come out in swarms in the rains, when some of the hillmen collect and eat them. The tiny fishes that I have mentioned, averaging perhaps an inch and a quarter in length, furnish little food ; and accordingly we have here no aquatic mammal and few birds to notice. The Three-toed Kingfisher (*Ceyx tridactyla*) is the most characteristic. *Halcyon leucocephalus* and *smyrnensis* occur, and probably the rare *H. pile ta* and *chloris*. *Alcedo bengalensis* is common ; and perhaps *Alcedo beavani* may be found hereafter.

The ubiquitous Paddy-bird and "Did-ye-do-it," and the smaller Sand-pipers, frequent the streams ; and the few and small tanks are used as resting-places by migrating Ducks and Teal.

The rivulets of the Konkan-Ghát-Mátá fall over the black cliffs of the Ghâts in innumerable cascades, separated by the terraces which run along the face of the mountains. Down to about 500 feet above sea-level there is no change in their population ; but here we find a tiny prawn associated with the loaches and *Discognathus modestus ;* and below this we come upon *Discognathus lamta* and a number of small Barbel and Carps, mostly, I suppose, fry of large species. Near the same level we begin to get a small Murrel ; and at the next step downwards the torrents unite to form small rivers, flowing through valleys of which the bottoms are usually under rice-cultivation.

These rivers very much resemble those of the Western Deccan described in my last paper ; but before they have time or space to unite and form important channels, they meet with the salt water. Probably no river of the Konkan has a perennial fresh-water stream fifty miles long.

There are however many deep potholes under falls ; and in some places long reaches of still water are formed by natural trap dykes crossing the streams or by artificial dams.

Some of the valleys are mere gorges ; others are of considerable width ; and these latter have usually flat bottoms, and appear to have been lakes within (geologically) recent times. Many of my readers are probably familiar with the theory that the basaltic floor of the Konkan, or at least of that part of it near Bombay, did, within the present period, sink westwards, somewhat as ice sinks from the shore when the water fails under it, immersing its western edge in the sea, and forming, amongst other things, Bombay Harbour, where there had probably been a lake surrounded by forest. In digging the Prince's Dock, a forest of Kheir trees (*Acacia catechu*) was found *in situ,* very much as you may see to-day the same trees growing in the forests of Mosare and Kirawli, five and twenty miles away ; and recent excavations in the salt marshes of Uran showed numerous roots and twigs with the bark on them : these however were not identified, and may have been mangroves ; but even this implies a depression of their bed, as mangroves do not grow below low-water mark.

The lacustrine remains found in the Island of Bombay itself may perhaps belong to another period. I am not personally acquainted with them.

But the recent depression that let the sea into Bombay Harbour would naturally spill the fresh water out of lakes lying further east, such, for instance, as the wide Panwell Basin, over which people look towards Bombay from the west edge of Matheran, or from the reversing station on the Bhor Ghât. The same thing probably happened to the valleys of the Kundlika and the Mangaum Kál. The line of disruption has never been exactly traced; but it is suggested that some clue to it may be obtained from the hot springs; and in that case it probably begins near Mhad, and runs through the valley of Mangaum and the very curious little defile of Ratwad, the Sukeli Pass, and the salt marshes east of the harbour; then between the Pars:k and Matheran Hills, and past Bhiwandi to Akloti on the Tansa River.

This however is all mere conjecture at present; and the main importance of the great break off to our subject is that it left us not a single lake in a country that was once probably a " lake region," and gave us instead estuaries in which the salt water often gets 30 miles from the sea. In some places on these creeks the mountains close in on the channel, and these defiles are often very picturesque.

But generally there is more or less flat salt marsh on one or both banks of each creek, sometimes reclaimed and converted into salt rice-land or salt-pans, but often covered with a dense growth of mangrove bushes, which grow to 25 or 30 feet high. The reclaimed lands are irredeemably ugly during eight months of the year; the mangrove swamps and islands, on the other hand, are very pretty at a distance or when the tide is in. At low water they are not pleasant neighbours from the heavy smell and hideous appearance of the bare mud about their roots, pierced by innumerable spiky and leafless suckers. The trees are not always true mangroves (*Rhizophoreæ*); indeed these are comparatively rare to the north of Bombay, but more abundant as you go down the coast southwards. The native name for them is *Kandel*, and they are easily distinguished by their strange flying buttress-like roots, glossy foliage, and flowers sometimes conspicuous and sweet-scented. Of this order, we have species of Rhizophora, Ceriops, Kandelia, and Bruguiera, and of others the " Tiwar" (*Avicennia tomentosa*) and " Surund" (*Excœcaria agallocha*), both of which are useful forage plants, " Phungali" (*Excœcaria majus*), with white flowers, and the strange " Marendi," or " Creek Holly," for which I have only a very old botanical name, *Acanthus ilicifolius*, probably superseded in late works.

The leaf is exactly like that of the common English Holly, and is sometimes used as a substitute for it in Christmas decorations, the *berries* being made up for the purpose of red beads cunningly tied on with wire. The flower is pale blue, rather conspicuous, with a superficial resemblance to that of a sweet pea. On embankments and other spots, raised ever so little above the marshes, we find the Chikhli (*Salvadora indica*), which so much resembles its relation, *Salvadora persica*, that one is surprised to find, apparently, a characteristic desert plant in so damp a situation. The fruit is of a much deeper and duller colour than in S. *persica*. For most of the description of these trees I am indebted to a report by Mr. Ebden, C.S.

As the estuaries near the sea, the salt marshes give way to clean sandy beaches in long bays, separated by promontories of trap-rock, and these beaches are generally backed by groves of cocoanut and other palms. The embouchure has almost always a steep and hilly. shore on one side (usually the south), and on the other a wide flat strand prolonged into a dangerous bar. Those of Bankot and Chaul are good examples. The smaller rivers which rise in the coast-ranges that run parallel to the Ghâts are miniatures of the larger streams that I have described ; but several of them debouch in the central part of flat plains, as, for instance, at Alibag and Warsoli. The plain here seems to have been once the bed of one of the lakes referred to above, the outer margin of which is still indicated by a line of reefs, of which Kennery Island and the Chaul Kadu Rock are the most elevated points. Subsequent to the immersion of most of the lake-bed in the sea, much of it has been reclaimed by the formation of sand dunes, originally backed by lagoons which have gradually become salt marshes. At this point the industry of man has stepped in to aid nature, and the sand dunes have become cocoanut gardens ; while the marshes, embanked so as to keep out the sea tides and retain the silt washed down from the hills, have become, first, salt rice-lands, and afterwards, as the silt accumulates to above spring-tide levels, capable of growing the superior rices which cannot endure even brackish water.

Wherever these reclamations have been made in creeks and backwaters, the mangrove swamps are of the greatest importance as protecting the water side of the embankments and furnishing materials for the repair of breaches. On the open coast, where the mangroves cannot face the surf, this function is performed by sand. dunes formed by wind and wave.. The total area of these reclaimed lands is very great, and their formation has within historic times. greatly changed the face of the Konkan. waters, and. must have seriously modified their population, expecially the Avifauna.

To seaward, immediately north and south of Bombay, that is from Dharavi to the Chaul Kadu Reef, the group of reefs, banks, and

islands of which Salsette is the largest and Bombay the centre, cover
a great number of sounds and inlets, mostly centring in Bombay
Harbour. Many of these are fast disappearing before natural silt and
artificial embankments, expecially the group west of Salsette and
that east of Hog Island and Karanja, both of which have been changed
from islands to peninsulas within living memory. This has given
rise to an idea that " the coast is rising ;" but if by this phrase
we understand an integral upheaval of the rocky sea-floor, there is
no evidence to support the doctrine. And in places where the coast
is directly exposed to the ocean alone, surveys made under my own
orders show that no change has taken place for nearly 30 years, that
is, since the first revenue survey.

The basaltic sea-floor, outside of the reefs and islands mentioned
(and from the coast itself north and south of them), descends by
gentle slopes, broken here and there by terraces, until at about 60
sea miles from the coast the " outer line of soundings" is marked by
depths, inside the line, usually of less than 100 fathoms, and outside
it in most cases of more than 200. This is a very rough description
of a matter deserving a fuller and better notice ; but for the purposes
of this paper, the " outer line of soundings" may be described as
marking a range of submarine " Ghâts" about 600 feet high, forming
the western face of a plateau continuous with the flat parts of the
coast and descending from it, by gentle slopes and small scarps, at the
rate of about 10 feet to a nautical mile. We know little positively of
its material, but are justified from its outlines and position in supposing this to be the Deccan trap, overlaid of course with marine
deposits.

The Orders, Genera, and even Species of aquatic animals which pass
from the salt to the fresh water are in places pretty numerous, and it is
therefore convenient to take the whole area together in noticing them.

The highest aquatic mammal of the Konkan is the Otter, which
inhabits all the creeks and streams and occasionally visits the sea, but
is not very common, and being a nocturnal beast and very shy is
seldom seen. It breeds in the hot weather.

After it come the cetaceans, of which we know but little. The
Indian Rorqual is known occasionally to visit the coast, and there
may be other large species. However, in a considerable experience of
the Konkan, I never saw a Whale spout in sight of shore but once.
It would be interesting if the experience of some of the officers of
the B. I. S. N. Company regarding this matter could be made

available. I have had two heaps of bones of Whales which had been stranded south of Bombay. One must have been over 40 feet long and the other under 30, so far as could be guessed from the *disjecta membra.* The latter was distinguished by possessing flat intervertebral plates of bone, which I could not find in the former. Neither had teeth.

Besides these, I have at different times received single vertebræ of at least two Whales. The last and largest of them is in the Society's Museum, and must have drifted a long way. It shows clearly the marks of the peculiar spades used by whalers in stripping off the blubber before " trying it out" into oil. But no whalers fish within many hundred miles of Bombay.[*]

We have at least two Porpoises—one a true Delphinus, called by the natives " Gadha ' (*i.e.,* Donkey), perhaps from his constant habit of kicking and frisking on the top of the water. There is a smaller one called " Bhulga," which is less common and is distinguished by having apparently *no buck fin.* It keeps in shallow salt water ; and I have not seen it frisk and play like the " Gadha."

I have never been able to get a specimen of either ;[†] they often get into fishing-nets, but almost invariably tear their way out. Some years ago some gentlemen from Bombay tried to harpoon them in

---

[*] There are Whale fisheries about the Maldives and Seychelles. The likeliest large Whale on this coast is *Balœnoptera indica,* the Indian Rorqual or Finback. I believe that a specimen in Bombay has been doubtfully identified as belonging to the allied genus Physalus. They have no teeth ; only whalebone strainers. Right Whales (Balœnæ), which have similar strainers and no back fin, are extra-tropical animals and need not be looked for here ; but the occurrence of a Sperm Whale or Cachalot (*Euphysetes simus*), with visible teeth in the lower jaw, concealed teeth in the upper, and a very small back fin, is possible, as of *Globicephalus indicus,* really a gigantic Dolphin, with a large back fin and visible teeth in both jaws.

[†] Since the text was written and sent in to press, I have received three specimens of the Bhulga, which has been identified as *Neomeris karachiensis,* and subjoin description, *viz,* old female, gravid; total length between perpendiculars, 4 feet 2 inches ; maximum girth, 2 feet 7 inches ; width of tail, 1 feet 8 inches ; length of flipper, 9 inches ; live weight, 60 lbs. avoirdupois ; colour, leaden black, lighter below, especially on the breast ; nose, chin, and interior of mouth dirty white.

No dorsal fin; but back behind the flippers flattened and hollowed out and carunculated; near the lumbar region edged with a slight salient angle, which may be taken to represent a rudimentary dorsal fin.

Mammæ 2, inguinal (of course), concealed in slit valves. No rostrum whatever. The profile rather reminds one of a Turtle's.

Teeth visible and numerous in both jaws (anxiety to preserve the specimens quickly prevented their being counted) in both adult and fœtus. In the former they are well worn down, showing that it is an old animal,

Spiracle crescent-shaped, single, central, and far back. No water was expelled from it in " blowing" during several hours that I had the animal under observation in water over its depth. I should say here that I am well acquainted with the Rorquals and *Globicephalidæ* in the wild state, and never saw either *spout water.* Their discharge is more like that of a starting locomotive steam-engine on a railway,

The contents of the stomach were many prawns (*palæmon*), mostly of large size, 3 to 5 inches long; three very small" bones"of sepias, the longest 2½ inches, and one pen of a squid (*loligo*) also very small. None showed any signs of dental action; they had apparently been swallowed whole. It is worth while to remark that the tongue of the "Bhulga," though distinct, is jaw-bound.

Mahim waters, but I believe failed. Two drifted fragments of skulls (from the Alibag Reefs) are in the Society's Museum. They appear to belong to different species.

Probably with a suitable steam-launch, and a combination of the rifle and harpoon, some very good sport could be had out of the " Sea-donkeys," which are extremdy numerous and not very shy. This has been tried with success in the English Channel. The sportsmen referred to above used canoes ; and I have tried to shoot them from a sailing boat, and (of course) believe I hit them. But I never bagged one. Of the Sirenia, sometimes called herbivorous cetaceans, *Halicore dugong* may occur, as it has been reported from Canara ; but our basaltic coasts are not rich enough in seaweed to feed it, so its appearance here is unlikely. It is sometimes called a " Seal;" but true Seals are seldom or never found between the tropics.

Of birds we have all those mentioned as found in the Deccan, and others more appropriate.

The chief of the marine raptores here is the Grey-backed Sea-eagle, called in Maratha " Khakan" (*Haliœtus lencogaster*). This bird is very common on the coast and creeks, and breeds here and there on trees. Sea-snakes seem to be the chief of his diet ; but he catches a good many fish too, and is said to rob the Osprey of his plunder. This I have not seen myself, though the Osprey too is common here, both on the salt and fresh waters, nor have I seen the Sea-eagle touch carrion or strike birds. He does not resort here to the fresh waters ; but the Osprey is seen on rivers and tanks as often as on the shore. The Brahminy Kite fishes a good deal on the surface of the fresh waters and creeks, seldom " out of harbour," and picks up carrion and crustacea on the shore ; and the Paria Kite (*Milvus govinda*) frequents harbours.

Some naturalists believe in a " large Paria Kite" (*Milvus major*) ; and Mr. Hume has recorded specimens from the dunes of Upper Sind and Bombay Harbour " which entitles him to a place here. To my own knowledge, there is in the forests of the Konkan a Kite answering to the description ; but whether he be really a separate species, or merely an aristocrat among " Paria " Kites, I don't pretend to say. The superior size and gentlemanly appearance of this bird, both on the wing and in hand, are very marked. The so-called " Blue Kite," or Harrier (*Circus swainsoni*), and Marsh Harrier (*Circus aeruginotus*), the White-eyed Buzzard (*Butastur teesa*), and probably the Long-legged Buzzard (*Buteo ferox*), hunt

NEOMERIS KURRACHIENSIS. (Murray.)

about rice-fields and the edges of swampy tanks and rivers for small birds, and probably for frogs; and so do both the Serpent-eagles (*Circaetus gallicus* and *Spilornis cheela*). I see that Lieutenant Barnes considers this last bird to be represented here by *Spilornis melanotis;* but I have shot many in the Konkan showing distinctly the marks which he insists on for S. *cheela, viz.,* conspicuous ocellation and barring on the lower surface and breast.* It is a common bird in the Konkan jungles. As with many other Eagles, the young of the year remain for some time with the old birds, and one can often hear three or four of them calling to each other out of trees or on the wing. It has several notes: the commonest is " Qui-yu-kuh," sometimes " Ku-qui-yu-kuh," " Kou-we-you" (rather long and deep), or a sharp repeated shriek " Qui-qui-qui." The Brown Fish-owl (*Ketupa ceylonensis*) is known, but being a shy nocturnal bird is not often seen. I never got a specimen myself.

Swallows can hardly be called aquatic birds; but it is worth while to notice that the " Edible-nest Swiftlet" (*Collocalia unicolor*) breeds in our present region on the Vingorla Rocks; and specimens of the nests from that place are in our museum. The theory that the nests are built of sea-weed, which would be a more legitimate excuse for bringing the bird in here by the neck and heels, cannot unluckily be maintained any longer.

The region is rich in Kingfishers, for which its streams are well fitted, being mostly well provided with small fish and overhanging rocks and branches.

Halcyon leucocephalus, the large Brown-headed Kingfisher, is rather common, and it is to me surprising that Lieutenant Barnes seems to think it a rare bird. It is tolerably familiar here; and I have often been able to watch one frequenting a tree near my tent for hours and days in succession. It has three notes at least. The common call is " Quí-yu-quí, Quí-yu, Quí-yu-quí." The alarm note is a harsh rattling laugh; and a wounded bird, when retrieved, has a "squawk" or " caw" very like that of a crow in the like case. *Halcyon smyrnensis* is common on all wooded torrents and tanks, and often at some distance from water, being largely insectivorous. The rare *Halcyon pileata* and *H. chloris* are both recorded by Mr. Vidal, and probably have escaped the notice of other observers, because on the wing, or at a distance, they were mistaken for *H. smyrnensis.* I have

* A Gujerat specimen shown at our September meeting as *S. cheela* has these markings, but less than many of my Konkan birds.

already mentioned *Ceyx tridactylus* as found in the Konkan-Ghât-Mátá; and as it is not essentially a bird of great elevations, we may be pretty sure that it exists on the better wooded streams below the Ghâts.

Alcedo bengalensis is very common on all fresh waters and on the coast, where it fishes in the pools left by the ebbing tide, and even in the surf on the reefs (not in heavy surf of course). One of these "long-shore" Kingfishers got to be very domestic in my verandah, which it frequently passed through on its way from the sea to a neighbouring tank, and would perch in for some time, taking refuge apparently from the violent rain-squalls which swept the coast. This was during the rains. The Blue Kingfishers seem to like sitting in the shade at midday in the hot weather; but *Halcyon smyrnensis* will also sit out on a look-out post, where he can see grasshoppers and the like. The Pied Kingfisher, on the contrary, seems to sit in the sun, because he likes it, and you may find him on every tank and open stream, on the creeks, and sometimes on the shore, where he is associated with *Alcedo bengalensis.*

The next set of water-frequenting birds are the Wagtails, which the natives call "Parit" (="Washerman"). They are rather numerous, and as a class well known; and their technical distinctions of this and that feather would be out of place here. They are on all fresh waters, and occasionally on creeks or even on the sea-shore.

The Weaver-birds, or "Bhayas," are water-birds in one sense, namely, that they almost always build near water and, if possible, over it. We have three species. *Ploceus bhaya* is common in the region. *P. manyar*, the Striped Weaver-bird, is more frequent at its northern end, where it opens into the plains of Gujerat, this being essentially a bird of the open country and of waters with reedy banks. *P. bengalensis*, the Black-throated Weaver-bird, is here rare and local; it has the same habits as *P. manyar*. Neither of the two last is as lively and interesting as the intelligent "Bhaya."

Of the Plovers proper, we have none of the Coursers, essentially moorland birds; nor, I think, any Swallow Plovers. The Grey Plover (*Squatarola helvetica*) is said to occur "all along the seaboard." I have never got it here myself, nor have I seen here, nor do I expect to see the Indian Golden Plover (*Charadrius fulvus*). If anywhere, these birds will be found on the occasional wide stretches of grass-land near the sea, such as the commons of the Alibag Taluka. Mr. Vidal has recorded the occurrence of the rare Caspian Plover (*Œgialitis asiatica*) ; and Lieutenant Barnes gives *Œ. geoffroyi, mongola,* and *cantiana* as coast-birds, and *Œ. dubia* and *minuta* generally for the Presidency. The last ought to be the *dubia.* It is a very dubious species indeed.

The European Lapwing is extra-tropical, and its nearest allies, the *Chettusiœ,* are rare cold-weather visitors here. Their place is taken by the Red-and-yellow-wattled Lapwings, or "Did-ye-do-its" (*Lobivanellus goensis* and *Sarciophorus bilobus*). The first is on every stream : the latter is less aquatic and rarer. The Stone-plovers *Œsacus recurvirostris* and *Œdicnemus scolopax* are not very common. The former deserves its name, frequenting sheet-rock and shingle in the beds of rivers and creeks (preferring fresh water). The latter ought to be called the Grass-plover, as its favourite quarters are in open grass-lands, and it is so independent of water as hardly to deserve a place here. It is the "Bastard Florican" of sportsmen, and does really seem by its habits to mark the connection between the Bustards and the Plovers, birds not widely separated by anatomical characters.

Of the *Hœmantopodidœ,* or Pied Sea-plovers, the Turnstone and Crab-plover may be looked for, and I think I have seen the latter. The Oyster-catcher is a permanent resident, and probably breeds here in small numbers.

There are absolutely no Wild Cranes in the region, probably because there are few cold-weather crops.

The Common and Pin-tailed Snipe are frequent cold-weather visitors, though the snipe-shooting of the Konkan is a poor affair to a man of Sind or Gujerat. The Pin-tailed appears to increase in number southward, which must be only an appearance, as both are undoubtedly immigrants from the north. The Jacksnipe is less common here than above Ghât ; they are all usually known as "Ishnáp ;" but the true Maratha name is "Shish." The Painted Snipe is a permanent resident, and breeds here in the rains, but has a curious habit of shifting its quarters in May, in small "wisps" of five to ten individuals, who are very careless of cover, perhaps because there is so little left them that they cannot afford to be particular.

Like the resident Ducks, the Painted Snipe is at this season fittest for the table, and no doubt for the same reasons as given in my last paper.

The Curlew remains on the coast all the year round ; but its little brother, the Whimbrel, seems to be only a cold-weather visitor, and is not so often seen, although the flocks are larger than those of Curlews. I have not myself seen the "Curlew-stint" on this coast. The genus (Tringa) seems to be chiefly represented by the little Stint (*T. minuta*), which appears in considerable numbers in the cold-weather. The Sand-pipers (*Actitis glareola*, *A. ochropus*, and *A. hypoleukos*) are common at the same season; the last less so than the two first. The Greenshank is common, and stays till April. The Red-shank comes in smaller numbers and for a shorter winter visit. The Spotted Redshank, if it occurs at all, is rare ; but the Little Green-shank is common throughout the winter, affecting fresh water and creeks rather than the sea-shore. The Stint is common on tanks rivers, and creeks. This bird and the Greenshank sometimes figure on butlers' bills as "*Woodykak*," for which they are very fair deputies. I have not seen the Avocet here. Most of the birds mentioned above go into the bag as "Snippets," or are contemptuously let off, which is a mistake in the case of most of them (unless Snipe happen to be plentiful), as they are good eating and quite as hard to kill on the wing as Snipe. Certain *shikaris* indeed include in their bags of "Snipe" pretty nearly everything that has a tolerably long beak. In one case I saw with mine eyes the murder of a Paddy-bird for the bag as a "Snipe" or "Plover ;" and indeed unless the term were pretty elastic, there would be no room round Bombay Harbour for the numerous sportsmen of the city. The firing there all Sunday morning in the cold-weather is enough to make one think the country up in arms.

The Bronze-winged Jacana is common wherever there are weedy tanks. Its ally, the Water-pheasant (*Hydrophasianus chirurgus*—why *chirurgus ?*) is much less so. I once saw one perched on a rock on the sea-shore. When disturbed, it flew off over the water to an island; but what brought it in such a place I cannot imagine. The Purple Coot is usually found associated with, or in the neighbourhood of, the two last birds, but is rare here. The Bald Coot is not very common. The tanks of the Konkan are too small for it. The White-breasted Water-hen is very common on the banks of rivers and in gardens, often at some distance from water. I have not seen the English Water-hen here. Water-rails and Crakes are not unfrequently shot amongst Snipe and Quail, especially by "griffins." I have no note of species observed.

Of Storks, the Great Adjutant and Jabiru (*Mycteria australis*) are rare. I have seen the former once below Ghât, and once in the

Konkan-Ghát-Mátá, to the best of my memory, and the latter only once below Ghat. The Black and the White (European) Stork I never saw in the Konkan at all. Here, as in the Deccan, the White-necked Stork (*Ciconia leucocephala*) takes the place and name of the former, and is pretty common.

The Herons are the same as in the Deccan, but far more numerous in individuals, especially in the creeks and salt marshes. Only the Purple or Grass Heron is uncommon, as there are few extensive waters with grassy banks. Most of them frequent the sea-shore : the exceptions are the Night Heron and (naturally) the Cattle Egret. Natives shoot the White Egrets (*Herodias*) a good deal for their dorsal plumes, which are marketable in Bombay. I do not think that any true Bittern occurs in the Konkan.

The Ibises, however, are pretty well represented by the Pelican Ibis and White Ibis ; the former on fresh waters ; the latter usually on the estuaries, where it associates with Curlews. The Shell Ibis is locally common on fresh water only. The Black Ibis (*Geronticus papillosus*) is rather rare ; and I have not seen the Glossy Ibis (*Falcinellus igneus*) at all. " Korle" is the Maratha name for both Ibises and Curlews.

The Spoonbill is decidedly rare. It is a bird of opener waters than we have here, where even the creeks are fringed (generally) with rock or mangrove. Now the Spoonbill does not like either rocks or trees. With it terminates the list of Fowl merely associated with water, and begins that of the Waterfowl proper. It leads up, in fact, to the Flamingo. In our last number I gave reasons for treating this bird as a Duck, and need not repeat them here. It is a migrant on the Konkan coast, but remains till June, in which month I have seen a flock flying north. I am not personally acquainted with the species or variety called by some writers *Phœnicopterus minor*.

No Swans and no true Wild Geese occur in the Konkan. The Black-backed Goose and its duodecimo edition, the so-called Cotton-teal (*Sarkidiornis melanonotus* and *Nettapus coromandelicus*), are found throughout the region, though both are rather uncommon. The climate suits them ; but the waters do not. They do not like salt water ; and the tanks and river-pools are not big enough for them ; but both may breed in favoured spots.

Of the next group of Ducks, the *Tadorninæ*, the Lesser Whistling-teal is found ; but it is not common ; and I have only seen it myself in the cold-weather· I was much surprised to find a small flock established on a rocky estuary, having always associated this bird, in my own mind, with grass and fresh-water. I have not found the Larger Whistling-teal here at all.

The " Brahminy " Duck is not common, and is even excluded from the Tanna District by the *Bombay Gazetteer*.

I have however once seen a pair in Bombay Harbour. Its relative, the true Shieldrake, has not yet been reported, I think, from the Konkan. We have the Shoveller, which is here a wild bird of respectable habits, and accordingly fit for the table. The European Mallard is unknown ; and its representative, the Spot-billed Duck, is not very common, nor, as far as my observation goes, a permanent resident. It is however extremely likely that when the Tansa Lake is filled, this and several other Indian Ducks will breed there. I hope that the Engineers will provide that lake with an island or two ; and that the Municipality will make it a sanctuary as regards birds. The shooting about Bombay would certainly be much improved by such a course, as Ducks like to make a large sheet of water their head-quarters, but will forage every day at considerable distances from home. The Gadwall occurs in the cold-weather, not in great numbers ; and the same is the case with the Pin-tailed Duck and Widgeon. The Common and Blue-winged Teal occur pretty frequently, especially the latter. The Red-crested and Red-headed Pochards are rare ; but the White-eyed Pochard is the most plentiful Duck on the coast and on creeks and tanks near the sea. I have not myself shot the Black-and-white-tufted Pochard here ; but I believe that I have seen it on the creeks, which are well suited to it.

Taking them altogether, the waters of the Konkan do not furnish good Duck-shooting. The birds mostly spend the day in the middle of the creeks, or on islands, or on the muddy and narrow margin between the water and the mangroves, where they are pretty safe that nothing can see or get at them from the shore-side at all, and nothing can surprise them from the water-side.

The deadliest way of killing them, no doubt, is to find out a feeding-ground in the salt marshes and lie in ambush (" flight-shooting," in short). But the pleasantest way of shooting on a creek is to take a boat or canoe capable of towing a small dinghy with one man in it, and run up or down the creek under easy sail and with the tide. The gunner is best placed in the bow of the boat, unless the sails be such as to interfere with him there. One man stands to each sail, and one to the painter of the dinghy, in which the "retriever" sits ready with his paddle or bamboo pole. Either of these is better than sculls, as the latter involve his rowing with his back to the game, or "backing water," and both manœuvres are inconvenient if he has to pursue a winged Duck.

Birds are not so much alarmed by the gliding motion of a boat under sail as by the more demonstrative processes of rowing or paddling, and will often give a sailing boat a shot. As the bird falls, the sail-trimmers instantly lower or brail up the sails, the man at the painter casts off the dinghy, and the "retriever" starts for his bird; while the helmsman brings his boat to the wind, or throws out a little grapnel or anchor; a stone does well enough. In the smooth creeks these manœuvres are not dangerous. When the retriever has got his bird (for which purpose he has, or should have, a light landing net) he rejoins the admiral, and the proceedings go on *da capo*. This is by no means a very killing way of shooting; but fair bags can be made, *plus* the poetry of motion in what is usually good scenery, and sometimes very beautiful indeed. Sometimes one should land from the boat, and employ her to divert the attention of the birds from a stalk, and this gives variety. The boat too enables one to indulge in a certain amount of comfort, and even, if necessary, to have books with one, to say nothing of fishing-tackle and belly-timber; and birds intended for preservation can be properly stowed away in a box or basket, or taken in hand at once. The rest of our water-birds are unfit for the table, or at least commonly thought to be. The first of them is the Dabchick, which is a permanent resident on tanks. It can however fly from one tank to another, and moves about a good deal more than it gets credit for, as it travels at night, probably for fear of Hawks and Eagles.

A "Mother Carey's Chicken" (*Oceanitis oceanica*) is known but rare. I do not know where it breeds; but on one occasion I noticed great crowds of various Sea-fowl near the Arabian Coast east of Aden; and the cliffs of that coast may well be the breeding-ground for some of our species. I don't know of any on our own coast.

I once got a live Shearwater, probably *Puffinus persicus*, which is in the Society's Museum. It was a storm-driven bird; and I have seen only one other in this region. I have not seen any Skua-gull here at all.

Indeed the poverty of this coast in Water-fowl is very remarkable to a fisherman trained on the Atlantic. Lieutenant Barnes speaks of the Lesser Herring-gull as occurring "in immense numbers all along the coast;" but I have never seen a really large flock of these birds here myself. Probably he referred more particularly to the Sind Coast. The Black-headed, Brown-headed, and Laughing Gulls occur, especially in the winter; the two latter go far up the creeks, and may sometimes be seen over rivers and tanks. The Sooty-gull (*Larus hemprichi*) occurs, but is not common. *Larus gelastes*, the Rosy-gull, may be looked for. I have not seen it here as yet.

Gulls indeed are much less numerous on this coast than the next group, the Terns. On account of the comparatively small area of permanent fresh water in the Konkan, the Marsh and River Terns are not very numerous; but we have in moderate numbers the Caspian, Gull-billed, and Whiskered Terns; and probably the large River Tern and Javan or Black-bellied Tern will be found hereafter, at least as stragglers.

A small Tern very common on creeks appears to be *Sterna minuta;* it may be Hume's *Sterna saundersi*, but I have a dislike to shooting these birds (which are very confiding, and often attach themselves to a boat and follow it for many hours), and cannot be sure of species not closely examined.

Thalasseus cristatus and *bengalensis* are common.

The Sooty-tern occurs, but is not very common. On inspecting after the south-west monsoon a beacon-tower on an exposed reef, I found in its chamber the remains, apparently, of a Sooty-tern, entangled with those of a banded Sea-snake about 20 inches long. It must be supposed that the Tern had caught the snake and carried him there to eat him, but been bitten by his victim, who was probably too much injured by the bird's beak to leave the spot. At the best, Sea-snakes are very slow movers out of water. I do not think however that our Gulls and Terns habitually attack Sea-snakes. Perhaps *Larus ichthyaëtus* may. Some of the large European Gulls would eat a baby if they found him unprotected. It was also very singular that the Tern should have carried his prey *inside* the tower. I can only account for the whole affair by supposing the bird was desperate from hunger in foul-weather.

The curious Skimmer (*Rhynchops*) does not occur here.

A white tropic-bird, or "Boatswain-bird," is not uncommon. It is probably *Phaeton candidus* or *Phaeton ætherius*. It gets the name of "Boatswain" from the fancied resemblance of its long pointed tail to a marlinspike, which (for the benefit of any reader that does not know) is a long thick iron pin, with a hole in one end, used for unlaying ropes. It is the characteristic tool of the boatswain, who is immediately in charge of all rigging ; and the proper place to carry it is in the back band of the trowsers.

We have one Gannet, or Booby, which is probably *Sula cyanops*. It is *not* Jerdon's White Booby (*S. piscator*), which has the bill and feet red, while in our bird the bill is slate-coloured, blackish towards the base, and the feet dull slate colour. It is a good deal to be regretted that recent naturalists have appropriated Jerdon's English name to a bird for which he certainly did not mean it, the more so because he prided himself on his system of English names, and took a great deal of trouble to make them clear and intelligible to everybody. A few birds of the present species are driven on to the coast every year by south-westerly gales, and are generally easily captured by hand. The present writer has sent specimens to the Society's and Victoria Museums, so there need be no doubt about the bird.

I don't think any Pelican occurs in the Konkan. If any, the grey species may be looked for, and there would be nothing surprising in its occurrence; but the fresh-waters of the region are rather too small for it ; and it does not seem to like sea water. The Large Cormorant does not, I think, occur ; and the Lesser Cormorant (*Graculus sinensis*) is not common. The Little Cormorant and Snake-bird are extremely common.

The highest reptiles of these waters are the Terrapins and Freshwater Turtles, which do not differ from those of the Deccan. I have never got their eggs;* but they seem to breed in the rains, as the young are very plentiful in October and November.

Two species of Sea-turtles are common on the coast. They are easily separated from those of the fresh-waters by having flippers instead of feet, and never showing more than two claws on a flipper, often only one.

The first is the Indian Green Turtle, *Chelonia virgata*, closely related to the Atlantic *Chelonia viridis*. The name I have adopted is sanctioned by its use in Dr. Gunther's " Reptiles of British India," and it is convenient to follow a standard work. It has thirteen shields of

---

* Since the text was written I obtained eggs of a Fresh-water Turtle (*Trionyx javanicus*) by dissection. They are almost spherical, cream-white, with a hard calcareous shell, about 25 in number. These Turtles, therefore, follow the Tortoises rather than the Sea-turtles in the matter of eggs.

tortoise-shell on the back, of a dull greenish-black colour; but the surface is always covered with little chips coming off, which give it a grey appearance when dry. When polished, it shows very pretty markings. These shields are no thicker than a sheet of thick note-paper, ; but the bony plates below them are sometimes as much as a quarter of an inch thick on the sides and half an inch on the shoulders.

The largest I ever got on this coast measured 5 feet between per-pendiculars, with his head as far in as he could withdraw it. The greatest total length may be taken at 5 feet 6 inches, and the live weight was 260lbs. avoirdupois. No doubt larger specimens occur; but from the information of a friend who had paid special attention to Turtles at the Nicobars (where they abound), I find that the average there is much the same as here; and anything over 160lbs. is a good Turtle. They are frequently caught in nets; and the females are surprised at night when laying their eggs. For this purpose they prefer mid-night and a spring high-tide, but are not strictly bound to time or tide; and I should not be surprised if they were found to lay in broad daylight on uninhabited coasts. They crawl up above high water-mark, often on grassy sand dunes several feet above it, and dig a hole, which is usually about 15 inches deep. The eggs, about 125 (but often far more numerous), are laid in the hole and covered with sand. They are at first of a very pale yellowish-pink colour, rather less than a racket-ball; and each egg has a crease in it. As development goes on, this disappears; the parchment-like skin of the egg becomes tight, and perhaps even stretches a little; at any rate the whole egg looks larger, and a dark blue stain appears on one side, the rest of the egg acquiring a dull white colour.

The Natives say that the old Turtle knows when the eggs will hatch, and then swims opposite the nest at high-water, and *whistles!* to the young, who, in obedience to the signal, tumble up out of the sand, and scuttle down to the water. The period of hatching varies greatly. The Natives put it at 3 weeks; and I know from experiment that this is sometimes enough. But I have now six clutches under observation, of which two are 42 days old and one 36 days. It depends upon the position and weather; shade and low temperatures evidently retard the hatching. Both wet and drought can prevent it altogether; the sand must be damp enough to keep the eggs cool, but well drained and, if possible, exposed to the full blaze of the sun. I keep most of the eggs in baskets, full of sand, set on bricks to secure drainage. The young are amusing creatures, very black and very active. If turned on their backs, they can right themselves like the Terrapins and Mud-turtles, and unlike their own parents. They are apparently omnivorous. At the time Dr. Gunther wrote, this species was supposed to live entirely on *algæ;* but if it could not do without these, there would be very few Green Turtles on this coast. The breeding goes on all the year round, chiefly, perhaps, in the autumn and beginning of the cold-weather. The eggs are just tolerable fried, or in an omelette.

The flesh resembles that of the "Alderman's Turtle" (*Chelonia viridis*), and is, of course, used like it for soup and cutlets ; but about the best thing to make of it is a kabob curry. It is said occasionally to be poisonous. If this is really the case, the cause is probably in some disease of the animal, and not in any natural changes ; for the most likely of these, exhaustion after laying eggs, certainly does not make the flesh of this Turtle unwholesome. The females, however, are naturally thin and poor at this period ; and the best meat is that of Turtles caught at sea, barren or not, far advanced in pregnancy, or males. The latter, I think, do not come ashore at all.

Our second Sea-turtle is the so-called Indian Logger-head (*Cawana olivacea*). It is not logger-headed nor olive-coloured at all, but has rather a fine profile—for a Turtle, and a good complexion, showing regular "tortoise-shell" colourings when wet. It seldom reaches 3 feet long ; it is less common than the Green Turtle ; and I have never got the eggs. It is reputed carnivorous ; and by some its flesh is thought inferior to that of the Green Turtle ; but I cannot myself make out any difference in taste.

Two other Turtles may be found here ; but I do not think that they are yet reported. The first is the Indian Hawk-bill or Tortoise-shell Turtle, which alone has · shields thick enough to make combs of. These overlap each other like the scales of a fish, whence the name (*Caretta squamata*). The other is the giant of the tribe, the Leather-backed Turtle (*Dermatochelys coriacea*), which has no tortoise-shell at all, but a thick skin laid over a ridge-and-furrow arrangement of bony plates.

The only Crocodile here is *C. Palustris.* I know that some specimens from Tulsi Lake have been exhibited at the Society's Rooms as *C. porosus;* but they all had the unmistakeable shields on the nape of the neck characteristic of the former species.

This is only locally abundant ; most so in the Kál River in the *Mangaum* Taluka of Kolaba. The fact is the fresh-waters and their

fish are not big enough for it ; and it is only an occasional visitor
to the estuaries, and very rare in the sea.

The Crocodile-shooting in the Kál is really good.

At Ashtami, on the estuary of the beautiful Kundlika River, there
is a small double-barrelled tank, containing innumerable frogs and
water-beetles, a very few tiny fish, and perhaps a score of Crocodiles over 5 feet long, besides youngsters, which keep in the
shallow water for fear of being eaten by their parents. They are
ludicrously tame. The oldest inhabitant had never heard of their
hurting any one ; and one could see them watching women washing
clothes, and mere babies paddling in the shallow water, without,
apparently, a thought of mischief. One over 6 feet long crawled
out within pistol-shot of my tent, and was shot by candle-light. He
had nothing in his stomach but water-beetles, may be a gallon of
them, and flint-stones swallowed, I suppose, to aid digestion.
A few days after a friend of mine (also a member of this Society)
wounded the patriarch of the tanks. A gang of life-boatmen, attached
during the fine season to my own and another private boat, dived
and literally harnessed him, and dragged him ashore, roaring,
snapping, and lashing the " scaly horror of his tail " like the old dragon. But on dissection we found the same water-beetles, *plus* two
crows which I had been using for bait in a vain endeavour to hook
him. This brute was 10 feet 2 inches long and over-pulled my spring
balance at 300lbs. I had watched them catching something all day for
ten days, and thought it must be frogs or tadpoles. The idea of such
brutes living entirely on water-beetles is new to me; and I would like
to know if any member has seen the like. Their teeth were quite
black, whether by reason of the water-beetle diet or not I cannot say.
Usually they are white, with brown stains.

Varanus dracœna, the Ghorpur, is very common, and eaten by
many castes. The name of Water-lizard is, however, misapplied to it.
In its habits it is a Land-lizard, which swims well, as many Land-
snakes do ; and can even dive well, which they generally cannot do.
But it is quite independent of the water, and is often found miles
from anything more than a well or puddle in a nullah.

Varanus lunatus, the Banded Ghorpur, may exist here. Young
Ghorpurs are all banded ; but this reptile is described as having 105
cross series of shields between the gular fold and the loin, as against
90 in the original Ghorpur. The Great *Hydrosauri* are not
found.

As for the Fresh-water Snakes proper (the *Homalopsidæ*), they are not, as a family, numerous here. This may be surprising to people accustomed to think of the Konkan as a damp and marshy country ; but the truth is that that description only applies to it for five months of the year. From November to May inclusive most of it is a very waterless country indeed to the great suffering of the people.

An estuarine species (*Cerberus rhynchops*) literally swarms in the creeks. As you sail up them you see a head popped up here and one there, and as instantly withdrawn, till you wonder what they all find to eat. It is an active reptile ashore as afloat, and the native name is *Udhan* (="the Jumper") from its peculiar way of springing forward. The Spotted Water-snake (*Tropidonotus quincunciatus*), which is not a true Water-snake but amphibious, derives from that nature a great advantage here and quite crowds out the Homalopsidæ. I strongly suspect that it fights, and even eats, them, but cannot propose to prove that just yet.

It has several varieties in colour, varying apparently with the colour and light of the water ; and ashore, it uses the same curious springing motion as the *Udhan*. It occasionally visits estuaries ; and I have taken small salt-water fish (*arius*) from the stomachs of individuals taken in nets in such places. So it is not a mere drift of the land-floods, but can forage in salt water. So does *T. punctulatus*.

These Fresh-water and Amphibious Snakes are not poisonous. The next family, the Sea-snakes, are all poisonous, though none of them can be called "deadly" in the same sense as the Cobra and Chain-viper, for a fair bite of whom there is no cure. Moreover, their fangs are very short, and a little clothing would guard a man from them. It is an additional reason for always wearing clothes when swimming in tropical waters, in some of which these reptiles swarm, if protection from the sun and from cold on landing be not enough to induce any reasonable man to swim in flannels. Except in racing, or at the moment of leaving the water, these are really no incumbrance at all, floating lighter than the human body.

Two genera of Sea-snakes, Platurus and Aipysurus, have the same classes of scales as Land and Fresh-water Snakes; that is, small scales above, and large ventral shields below, the latter acting as feet. I believe that neither genus is represented on our coast. If anywhere, they should be looked for on shores and in marshes, for we may be quite sure that the ventral shields exist in them, as in terrestrial

Snakes and the Homalopsidœ, to enable them to move on land, or
at least on mud. They are, in fact, Shore-snakes rather than Sea-
snakes, though, like the Fresh-water Snakes, their nostrils are placed
high on the snout, and, like the Sea-snakes proper, they have, in
addition to this, the ventral region more or less compressed; and the
tail flattened out into an oar, to be used as the single and sculling oar
is in a merchantman's dingy. This motion however is not in any
Water-snake or in any Land-snake (swimming for the time) confined to
the tail. The undulation of the whole body propels it forward, and in
some of the most essentially marine species the flat tail, properly so
called, is insignificant; and the abdominal region does most of the
propulsion. Snakes, in fact, move in water, as on land, by undulation.
Only in the former medium, their best purchase is on their two sides.
On shore, it is naturally on the belly. True Sea-snakes, stranded,
are even more helpless than fish in a similar position, for the latter do
then use their lateral fins on the bottom as legs, and often regain
deep water in that way.

But the Sea-snakes, with their lax bellies and small scales, lie
helpless. They wriggle truly, but on one spot, like a rocking-
horse; and they generally remain till a passing man squashes their
heads, or a bird of prey carries them off. The Grey Sea-eagle is
a great hand at this, and always goes once up and down his beat
on the coast, every tide, with a view to tide-falls of the sort. These
Sea-snakes without ventral shields, mostly belonging to the genus
Hydrophis, are of a great many species, and offer considerable
variety in form. I might almost say that amongst them there are
analogues of most venomous Land-snakes. With a single exception,
however, they are of very similar colouring, banded black and white.

The bands take different shades. In some they almost merge in
a general dull grey; in some the light favouring, you can call them
purple and yellow. They are continuous round the body or forked,
a single band on the right side meeting two from the left, or those of
each side alternate; but the type is general.

There are exceptions to it. One is a very widely distributed Snake
(*Pelamis bicolor*), which has several varieties. That commonest here
is, when young, velvet black above, on the abdomen golden yellow, and
on the flattened tail handsomely mottled black and white above and
below. As it ages, apparently, these brilliant colours fade to a dirty
olive on the back, and equally dull white below, all over; but one
specimen which I have sent to the Society's Museum seems to

have retained its colours to maturity. Another is the new *Hydrophis phipsoni*, striped black, white, or grey.

The *Pelamis* is the only Sea-snake that justifies the Ancient Mariner's description. The rest are loathsome reptiles. In many hundred specimens I have not witnessed the ferocity ascribed to them by Dr. Gunther. In one case only I saw one bite itself, apparently with no ill effects, though the species (a Hydrophis) was certainly venomous. They are held in great contempt by the fishermen, though these well know their poisonous qualities. On one occasion, being in the water with half-a-dozen naked men, I saw a Hydrophis, 4 feet long, swim towards us, and called to a man who had a bamboo to kill it for fear of accident. He did not hear me ; but a naked man, who did, picked up the reptile in the most unconcerned way, and chucked it on to the sand, where it lay helpless.

Pelamis is much more active both afloat and ashore, and gets more respect accordingly.

My fishermen call all Sea-snakes " Kilis " in Maratha.

For the Great Sea-serpent, we know nothing of him here, except that he cannot be of any type of Sea-snake known to us. For, if he were Platuroid with ventral shields, he would surely come ashore to exercise them; and if he were a true clumsy Hydrophis without ventral shields, he would as certainly get cast ashore sometimes, as that tribe and the whales do. Or at least an odd bone would drift to us, as my bone of the " whaled " whale did from unknown, but certainly very distant, regions, with the cuts of the blubber-spades on it. The bones of Sea-snakes float easily.

Our Sea-serpents do not often reach 6 feet long, but we read of their attaining 10 feet.

It does not follow of course that there cannot be a Great Sea-serpent of a totally different type, possibly far more saurian or more fish-like.

KESWAL.

BIRD-NESTING ON THE GHATS.

By Mr. J. Davidson, C.S.

I HAD paid a short visit to the Kondabhari Ghât in August 1885, and the beauty of the place at that season, and the number of birds evidently breeding there, made me determine to go there this year at an earlier period, when I would find fewer young birds and more eggs.

My transfer however to another district seemed at first to make this impossible. Thanks however to good early rain, the population were too much taken up with their farming to quarrel with their neighbours, and I found I could get away for a week without any great inconvenience to any one.

It was therefore with a light heart that on the afternoon of Saturday, July 10th, I left my head-quarters on a week's casual leave *en route* to the Ghâts.

A rapid drive of some thirty odd miles brought me to Dhulia, the head-quarters of Khandesh, in time for dinner, and I was fortunate enough to escape without any rain, though the country near Dhulia was almost under water, and I could see heavy rain following me nearly all the way.

The crops were looking well as long as it was light enough to see them; but bird-life was not abundant, and all that I saw worthy of notice was a solitary Adjutant (*L. argalus*) accompanying some Grey Cliff-vultures in a banquet on a dead cow.

The Adjutant is never common here, and during the five or six years I have known these districts, I have not seen a dozen in all, and always single specimens, and that during the rains and cold-weather. The Adjutants in the east of India seem mostly to resort to Burmah for breeding, and breed there in October; but no one seems to have found out where the birds from Western India breed or when.

A little further on I saw a Roller (*C. indica*, not *C. garrula*). This was distinctly exceptional. During the cold-weather *indica* is very common everywhere in Khandesh and Nasik; but in the hot-weather it leaves the plains and breeds abundantly in the Satpuras and Ghâts, and at the beginning of the rains it appears to leave the district (plains and hills alike). From the beginning of June till the middle of August one hardly ever sees a Roller. About that date, *C. garrula* appears about Dhulia in some numbers and remains till October, in the beginning of which month and the end of September *C. indica* also returns. Sunday, the 11th, I spent in hospitable Dhulia, and the juvenile Bhil population as usual brought a variety of nests and eggs. These consisted of the usual common Dhulia birds— *Priniae* (*hodgsoni* or *gracilis*) and *Stewarti* (for the first two birds are one species), *Franklinia buchanani, Pericrocotus peregrinus, Caprimulgus asiaticus, Drymœca inornata*, and *Sylvatica*, &c., &c., the only nest requiring special notice being one of *Volvocivora sykesi*. This pretty little Cuckoo Shrike is one of the earliest migrants in the rains, arriving about the 8th of June, and breeding all along the scrub-jungles which stretch between the Nasik and Khandesh Collectorates. It appears particularly partial to the Angan forest, and, as far as I remember, all the many nests I have seen have been in forks of Angan trees. The nest is a pretty firm platform, composed of fine roots ; and the eggs, which much resemble those of the Magpie Robin, are three in number.

The only bird I noticed specially at Dhulia was a single Alpine Swift (*C. melba*). In that most useful book Barnes's "Hand-book of Birds of Bombay," he states that this bird only occurs as a somewhat rare cold-weather visitant. In this I think he is mistaken, and that *C. melba* is a permanent resident in all parts of the country where there are high enough cliffs to afford safe breeding-places. I have been told that it breeds in Kanara at the Gairsoppa Falls ; and I find in my note-book records of having seen it in Nasik and Khandesh in every month except October and November, so have no doubt that in this part of the country it is found throughout the year. Last May I saw flocks of hundreds flying into and out of fissures in the cliffs at Saptashring near Nasik, and though I could not get near the places, I have no doubt they were then preparing to breed. If they breed there, their presence anywhere within 200 miles would be nothing extraordinary, judging from the pace they fly at.

In the afternoon, about 4 o'clock, I left Dhulia and drove due west to Sakri, 33 miles, noticing on the road another Roller, apparently also *indica*. As it grew dark, occasionally a pair of Painted Sand-grouse passed across the road, and the cries of many Nightjars (mostly *asiaticus*, the others *monticolus*) were heard on every side. These birds are all common inhabitants of the scrub-jungle here which adjoined the road on each side.

I rose early on the morning of the 12th and by 8 o'clock had reached my destination, a rather dilapidated bungalow or tool-shed, belonging to the P. W. D., situated on the edge of the pass. On the road great numbers of males of a pure Yellow Moth (a *bombyx* apparently) were flying about. I had however no net with me, and did not attempt to catch them.

On arriving at Kondabhari I at once took a short stroll down the Ghât. The place is an admirable one for bird-nesting. The hills in the neighbourhood are very steep and slope down on the Nowapur Pergunnah, a sort of northern edition of the Daugs, with the same unhealthy climate and water. At this pass a small stream runs down and forms a valley seven miles long and in no place at all steep.

Government have constructed a very fair road down the valley, and as the hills on both sides are densely wooded, it is both a capital place for birds and easily worked.

I wandered down the nullah for a mile or so and found lots of birds; but nests were few and far between, and when I got back, about 11 or 12, o'clock, all I had found were some dozen of the beautiful hanging nests of *Zosterops palpebrosa*. Most of these were empty; but three or four contained young of various sizes, and two had each four eggs—in one case fresh, in the other unblowable. The nests were in every case suspended over the river (then dry), and varied in height from the ground from 7 to 20 feet. I also found four nests of *Myagra azurea*—one with a fresh egg, which I left, and the rest either empty and old or with big young. This bird is very common on this Ghât, and makes its nest, generally on an " Umar" tree. It is a very beautiful structure—a deep cup, generally attached to the side of a single hanging twig. Its sides are beautifully ornamented with white nests of some spider, the pattern being so regular in some cases as to resemble. lace-work. I noticed a single pair of *Muscipeta paradisi* in chesnut plumage. They are rare at this season here, and I watched them a long time but saw no signs of their breeding, and when I again visited the place a couple of days later they were gone.

In the evening I again went out and worked up the nullah. In the first few paces a pretty little Blue Robin (*C. tickelli*) darted from its nest. This was placed in a crevice of the bank, and might have been mistaken for one of our own familiar Robin Redbreasts. It contained three olive eggs, perfectly fresh. The Blue Robin is one of the commonest birds at this season along the Ghâts, and its pretty metallic song seems never to cease if you wander along any of the nullahs. Its nests, of which I found many, including four or five with eggs, were placed in hollows either in banks or in the roots of trees, and were composed of dead leaves, lined with fine roots, sometimes intertwined with hair. I had hardly packed these eggs in my box when one of the Bhil boys noticed a large rough nest on a bare tree close to the nullah. It was a difficult tree to climb, and the boy declared it was an old one, but was promptly sent up to make sure. He scrambled unwillingly up, and as his hand was touching the nest, and his tongue again pronouncing the antiquity of the structure, a short-tailed bluish bird darted out. This was a specimen of the beautiful Yellow-breasted Ground-thrush (*Pitta brachyura*), and the nest, which was a clumsy structure of fine twigs, lined with dead leaves, contained five slightly-set eggs. They were almost round, of a beautiful China white, with dark magenta blotches and lines scattered over them.

A few hundred yards further on two similar nests were found—one empty, and one containing five fresh eggs. The stupid boy however broke one in bringing it down.

A heavy shower of rain now came on, and in the narrow gorge we were in it was too dark to see anything, and we were fairly driven in.

The morning of the 13th was fine, and 1 drove a couple of miles down the pass and searched all the jungles on the left side of the road downwards. Birds were numerous, and I obtained two nests each with four eggs of *C. tickelli*, two nests, with one and five eggs respectively, of *Pitta brachyura*, as well as two empty nests of the same bird, one of *Alcippe poiocephala* with three eggs, and one of *Myagra azurea*, also with three eggs, and one or two of *Zosterops palebrosa*.

All these birds were noticed again and again, though *Alcippe poiocephala* is much commoner 50 or 60 miles further south. One of the Bhils also knocked over with a stone a fine specimen of the Rufous Scops Owl (*S. sunia*), if it is really distinct from *S. pennata* It is a full grown male, and only measured 6·1 in length. I noticed many specimens of *Scops bakhamuna*, the Grey Scops Owl of this district and the Satpuras. They are however very much larger birds, measuring from 8 to 9 inches. The Rufous Owl I have only found in this Ghât and during the rains. The evening I devoted to endeavouring to watch specimens of *Parus nipalensis* and *Machlolophus aplonotus* to their nests. The former was very common; the latter scarcer. I watched both pairs and single birds ; but in that thick green jungle I invariably lost sight of them in some thick tree, and whether they had entered a hole or merely flown on to another tree I am to this day no wiser : I certainly found no nests.

On the morning of the 14th I took a lot of Bhils and walked down the nullah, taking the same ground I had gone over on the Monday. This is really much the best part of the jungle, and I was disappointed at its barrenness. I took the eggs, now three in number, from the nest of *M. azurea*, and got also a couple of nests of C. *tickelli*. In one however the eggs were ready to hatch. I did not disturb them, and the other was only building. I also found a nest containing three nearly full-grown young and one addled egg of *G. cyanotis*. This Thrush is not common here, and this seems in this part of the country about its northern limit, and it is only a migrant, arriving in the rains; it is however common enough

along the ridge running eastward from the Ghâts immediately north
of Nasik.   Thrushes as a rule are very rare in Khandesh.   I have
only seen one specimen of *M. horsfieldi* and one of *M. nigropilea*,
while the former is fairly distributed, though rare, in the north of
Nasik, 40 miles south of this, and the latter simply swarms along the
Saptashring Range stretching from the Ghâts eastward ; so much is
this the case that in a week the patel of one village sent me in 70
eggs of this bird collected in one small hill.

I watched a pair of Jerdon's Green Bulbul (*Ph. jerdoni*) for a long
time, but they had evidently not commenced to build, and I shot the
cock.   Last year I obtained nests with eggs in this same Ghât in
August.   I also shot a pair of Indian Cuckoos (*C. micropterus*), or,
more strictly speaking, two specimens, for they were both cocks.

Every day I heard a clear note I could not make out ; and finally
I followed it up and shot these two birds in the act of calling.   It is
not very common, and this is the only place I have noticed it in
Khandesh.   All the time I was at the Ghât I never saw or heard the
European Cuckoo (*Cuculus canorus*).   This is a very similar bird, but
the much narrower bars on the breast make it very easily identified.
It passes through Dhulia in the early part of June, and in July is very
common throughout the Satpuras, a dozen often being heard at one
time.   It returns again in September, and no doubt breeds abundantly
in the interval.   As the Satpuras are barely 50 miles north of these
hills, it is strange none of the Cuckoos stop to breed here in the rains.

Coming back I got a nest with three eggs of *Leucocerca leucogaster*.
This pretty little Fantail is very common on the Ghât ; but its nests are
difficult to find, and the bird was not rare enough to make me willing
to waste time over it.   I only noticed one pair of the larger kind
(*Leucocerca aureola*), and that was well down the Ghât.   It however is
common on the plains above.   I noticed one Honey Buzzard (*P. ptilorhynchus*) ; and the shrill cry of the Hawk-eagle (*L. cirrhatus*) was
constantly heard.   This bird is common here and in the Satpuras ; but
in the adjoining parts of Nasik I have never noticed a single specimen, and it is far too noisy to be passed over.   In the evening the
villagers brought me a number of Mynas' eggs.   These must have been
from second nests, as there were lots of young flying about.   All were
the common species (*A. tristis*).   Indeed I have never seen a single specimen of the Blue-eyed Jungle-myna (*A. fuscus*) in Khandesh, though
it is common on the hills immediately south of that district.

On the 15th I had determined to have a day in the jungles at the foot of the Ghât, and had sent the Bhil boys down the night before. The morning was however very wet, and it was past eight before I started. It was still wetter when I reached my destination, and I was glad to take refuge in a dharamshalla. About half-past ten it looked a little clearer and I ventured out, and by twelve it was quite fine. Everything was however soaking wet, and naturally I got very little. I found one nest of *Dumetia albogularis* with four fresh eggs. This bird I found in great numbers last year; but it was much scarcer this year, and I only got one other nest with hard-set eggs. The nests are placed on the ground, and are quite round, composed of long dry grass, the entrance being at the side. With the exception of a nest, with one egg, of *Alcippe poiocephala*, all the nests I found on this occasion were of common birds, and there were fewer birds and nests at the foot of the Ghât than along the sides of the nullah higher up.

The 16th was my last day, and I walked along the road for a couple of miles and then took the other side of the valley. I twice heard the mournful wail of a Ground-thrush. The bird's cry in the breeding season resembles that of a young Spotted-owl, and no one would ever dream it was the cry of a Thrush. I found the nest of one pair, but the eggs were not yet laid. In the other case I could find no nest, though the birds kept flying round and round me, and I think I examined every possible tree. The neighbourhood of this second nest (for there must have been one) was a very good place for nests. Within 100 yards I saw a brood of *Buchanga cærulescens* just able to fly, and also one of *Oriolus melanocephalus*, while on an adjoining tree there was another nest of this Oriole with two slightly-set eggs. It was a very deep cup on the end of a thin branch, and though in cutting the branch to get at the nest it got turned at right angles to its proper position, the eggs were uninjured. I do not think this nest belonged to the same pair as that which had young ones flying.

These Orioles are very common here, and I found four nests : one was new and empty ; from another the birds had just flown ; while the remaining one contained one fresh egg. The bird would no doubt have laid more ; but to get at the nest I had to cut the branch off, and it was only then I discovered that only one egg had been laid.

On the very next tree to the one with this bird's nest was an empty Thrush's, and 20 yards off a nest of the Common Ghât-babbler (*M. malabaricus*). This bird never seems to leave the jungles, and as soon as cultivation begins on the top of the Ghât is replaced by *Argya malcolmi*. The nest in question contained one fully-fledged young Hawk-cuckoo (*Hierococcyx varius*), and there were *three* old Babblers in attendance. Nothing could induce the Cuckoo to leave the nest; and finally the boy threw nest and all down and it still held on with its claaws and bit at my finger. *H. varius* is very common in the valley and on the hills above it, and its shrill cry " Pu-pe-ha," " pu-pe-ha," re-echoes from every hill, and not one Babbler's nest brought to me was without one of the round eggs of this bird. In each case the embryo Cuckoo was much further advanced than the Babblers; so it seems certain that the Cuckoo lays its egg before the Babbler does, or that its period of incubation is less. Proba-bly both are the case, as I remember once starting a Hawk-cuckoo out of a small bush and finding in it a nest of *M. malabaricus* with a fresh Cuckoo's egg, but no Babblers' eggs. I have litle doubt the bird I disturbed had just laid this egg; but as *Coccystes jacobinus*, which lays very similar eggs, is common in the Satpuras where this occurred I cannot be absolutely sure. During this visit to Kondabhari I neither saw nor heard *Coccystes jacobinus*, so am sure that the Cuckoo's eggs I obtained all belonged to *H. varius*.

A pair of *Graculus macei* was also apparently breeding near this place. I could not however discover the nest, though I watched for a long time. I found a nest with two young in September last near the same place; but in thick jungle it is easy to overlook a nest placed high up in a fork and of exactly the same colour as the bark.

On my return to the bungalow I found, among other eggs collected for me, two nests, each containing five eggs, of *D. sylvatica* and *Cisticola cursitans*. The former is moderately common both on the Ghât and in the Maidan above, but the latter never seems to enter the high grass and jungle of the valley.

Afterwards I had to leave and drive back to Sakri. On the whole I found much fewer birds than I expected; but of course no migratory birds had arrived, and few young birds were flying. The only mam-mal I saw was a Hare. There were fresh tracks of two Panthers and a Hyena and a few four-horned Antelope, but I never came across any of the animals themselves.

Moths were common, but Butterflies were few. Of the *Papilionidae*, the only one really common was *Eratonius*. I noticed a few specimens of *Polites*, *Agamemnon*, and *Nomius :* the first two were fresh from the chrysalis, and the last very battered. Caterpillars were however abundant, and included several of the *Sphingidae* and two beautiful Green Caterpillars of *Actias selene*.

On the morning of the 18th I returned to Dhulia, where the nests of a great many common birds were brought to me ; and on the 19th I returned to my station and a vast pile of arrears of work. On the return journey the only thing I noticed was the number of males of *Pericrocotus erythropygius* which were flying about as we passed through the angan jungle near Arvee. This very handsome bird is very common in this narrow belt of jungle, and the hens were evidently sitting. In previous years I have taken many of their nests, the restlessness of the birds, who are constantly flying to and from the nest, at once betraying its position.

J. D.

NOTE ON SOME POST-PLIOCENE MOLLUSCS FROM THE BYCULLA FLATS.

COMMUNICATED BY MRS. W. E. HART.

You will doubtless smile if I speak to you of the treasures of the Byculla Flats. But I can assure you the whole of that much-abused region is full of interest, both for the geologist and the zoologist. At no very distant (geological) date the sea must have ebbed and flowed freely eastward, past the site of the Byculla Club, over all that ground now covered with cotton mills and municipal refuse heaps, and intersected by causeway roads and open sewers, which lies between Mahim to the north and Cumballa Hill to the south. At a comparatively recent date its approach from the west was in some degree barred by a sort of breakwater formed by the elevation of the coast-line at Worli. This is clearly shown by the occurrence of fragments of modern sea-shells in the red earth at Worli Point, 16 feet above the present high-water mark. The sandy isthmus just south of Worli village too, in which modern sea-shells are found in good preservation and in considerable quantities 6 feet above the present level of high-water, must once have been the sea-beach at this point. But it was not till the construction of the Vellard Causeway at Mahaluxumi by an English Governor of Bombay in modern times that all access to the Byculla Flats was finally denied to the sea. In the interval it still continued to enter from the south-west at the indentation south of Love Grove, and spread in a broad shallow lagoon over the present level of the Byculla Flats. This was slowly filled as the tide flowed in through the narrow opening between Mahaluxumi Point and the southern extremity of the Worli Ridge, and again was emptied, or nearly so, as the tide flowed out, much in the same way as we see to-day the low-lying ground about the muddy creeks of Salsette. The entrance being so small and the space beyond so great, it is clear the tide can never have flowed with any great force over the slowly shelving ground inside. Hence it is natural that whatever creatures died in these sheltered shallows, or were drifted into them from the sea outside, would there soon be silted up and preserved in the soft in-washed mud. Hence it is that you will find every spadeful of the soil of the Byculla Flats literally full of the remains of countless sea creatures in a semi-fossilized condition, and for the most part in a wonderfully well-preserved state.

This lump of earth, marked No. 1, is an instance. It was found near the race-course on the top of a bank of earth made of the soil excavated on the spot. Of course the texture of the shells in many instances is greatly altered, or even completely changed, generally owing to the highly aluminous nature of the clay or siliceous condition of the water in which they were deposited. But this only shows how long such specimens must have been lying undisturbed exposed to these influences ; and the fact that they have so well retained their original forms shows how very gradual was the operation of the influences to which they were subjected. The group of fragments of tubular shells, marked No. 2, illustrates this alteration in texture while the original form is preserved. A yet more curious illustration is afforded in every handful of earth about the brick-fields on the west of the Byculla Flats. The soil here is somewhat laminate, very friable, and full of small crystals, apparently of gypsum. And its effect on the shells buried in it seems to have been in some instances to crystallize them, and in others to turn them a dark brown or black colour. In either condition they still retain their original form, but are so brittle, or rather rotten, that the slightest touch reduces them to powder, and I have found it impossible to bring any here in a recognisable shape.

The alteration in texture, considered in connection with the nature of the surrounding soil and general character of the locality, would be of special interest to the geologist. But the shells themselves, whether their texture is altered or not, present several points of interest to the zoologist which I venture to think would amply repay their careful study by a skilled conchologist. I have therefore presumed to invite to them the attention of the members of the Bombay

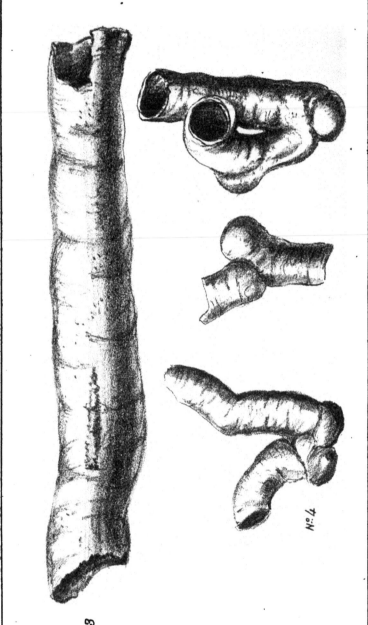

N°3

N°4

R.A.S del.

A.C. GARTEL. LITHO

Natural History Society in the section of " *Other invertebrata*" in the hope that some one may be found more competent than myself to discuss their characteristics.

It seems to me such a discussion might be specially profitable in two ways. First, a careful comparison of these sub-fossil shells of the Byculla Flats with those at present inhabiting the sea outside, with a view to determining such slight differences as may have become permanent during recent geological ages, might throw much light on the theory of evolution and the origin of species ; and secondly, from a study of these marine remains on dry land, we may learn so much of like creatures still inhabiting the sea as to be enabled more easily to find living specimens of species hitherto considered rare from their inaccessibility. It is chiefly in regard to the latter consideration that I propose to offer a few remarks now on these fragments of tubular shells which I have produced for your inspection here to-day.

Among the commonest of the shells scattered over the Byculla Flats are some not unlike pieces of the broken stem of a clay tobacco-pipe. My attention was first directed to them about two years ago by Major R. T. Frere, R.E., who believed them to be the tubes of some boring mollusc. Unfortunately he was compelled by ill-health to go to England before he had prosecuted his researches very far. He took with him however some specimens he had found, and later I sent him some I found after his departure. By comparing these with specimens in the collections of the British Museum and the Royal College of Surgeons, and by the help of information and assistance courteously afforded him by the officials at these two institutions, and particularly by Mr. Etheridge, the head of the Palæontological Department at the British Museum, he collected some interesting information regarding the natural history of tube-forming animals. This he has kindly imparted to me, and I beg to lay before the Society such portions of it as seem to bear upon the specimens which I have collected from the Byculla Flats.

When found in their least altered condition, the tubes are apparently calcareous and nearly white in colour, or faintly tinged with pink. They vary considerably in size. But I have found no fragments larger than those in the group marked No. 3 either in point of length or circumference. I think the reason of this is that the creatures inhabiting these shells used to bore downwards into the soft oozy bottom of the lagoon I have described, big end first. The

excavations hitherto made on the Flats have not yet passed the level
at which they attained this circumference, and the length represents
the extreme diameter of a clod ordinarily loosened by a stroke of the
pick in the work of excavation.   When this is lifted, of course so
much of the tube as it contains is snapped off and carried away with
it.   I daresay if we were to dig carefully downwards from the
present level of the Flats, we should be able to uncover specimens of
greater length gradually increasing in circumference, till at last we
reached the lower or big end, as to which I shall have something to
say presently, but a specimen of which I have not yet succeeded in
finding.

One curious characteristic of these tubes is the way in which they
change their direction, as shown in the group of specimens marked
No. 4.   I would particularly draw your attention to a feature in these
to which I shall have occasion to refer again, *viz.*, that wherever
one of these changes in direction occurs, it is marked by a little
rounded knob or excrescence on the shell.   Probably these changes
in direction were necessitated, either by the inhabitant of the shell
coming on some hard substance through which he could not bore, or
by his being obliged to work in a very confined space, by reason of
his neighbours crowding on him, or by reason of the limited extent of
the soil suitable for his operations at the scene of his labours.   The
excrescences, I presume, were formed by the animal closing the end of
his tube in the old direction when he started in the new to prevent the entry in his rear of water or mud or animals which might
cut off his connection with his upper or smaller end.

What then are these tubes?   Before attempting a solution of
that question, it may be as well to state what they most certainly are
not.   They are not calcareous casts of the stems and roots of aquatic
plants, formed by the deposit of lime held in suspension by the water
in which they grew on vegetable substances which have since decayed,
leaving only their mineral envelopes.   You may think that in
enunciating such a theory for the mere purpose of demolishing it I am
but setting up a man of straw for the pleasure of knocking him
down.   But I remember Major Frere once telling me of a passage in
some work on the geology of Bombay, in which it was suggested that
the shelly tubes found on the Byculla Flats were casts of the roots of
the mangrove bushes once growing there when the place was a muddy
salt marsh.   I have forgotten the name of the book, and I have not
been able to find it since ; but I believe it came from the library of the

Bombay Branch of the Royal Asiatic Society.* There was much in it that was interesting and valuable. But this theory was certainly wrong. By a curious coincidence, I happened only the other day to pick up near the race-course this specimen, No. 5, which shows, I think, how the theory of the mangrove roots may have originated. You see here is a bit of the root end of some plant firmly embedded in a fragment of tube. It may either have grown up naturally through the tube, which had accidentally fallen and become embedded in the earth in such a position as to allow of this, or it may have been poked in to clear the tube by some inquisitive cooly five minutes before I found it. I incline myself to the former theory, both as the more interesting, and because when I first found the specimen, it was completely filled up to the edge at both ends with earth, some of which has since been shaken out, which would hardly have been the case had the piece of stick been pushed in for the purpose of cleaning out the earth.† But however it got there, there is the piece of the plant in the piece of the tube, and it is not impossible that a hasty observer might jump to the conclusion that the latter was deposited round the former in the manner suggested by the author of the work to which I have referred. The general objection to the theory is that the fragments of calcareous tubes are always single, whereas the roots and stems of mangroves are always branching. In this special instance the space between the shell and the wood, now filled up with earth, shows that the former can never have been deposited on the latter.

But after thus disposing of the theory of another, it is only fair that I should give him a chance by advancing one of my own. My theory is that many (I admit not all) of the shelly tubes found on the Byculla Flats are fragments of the tubes of an interesting and hitherto rare mollusc, belonging to the family of Pholadidæ, and variously known as *Kuphus*, or *Septaria*, or *Furcella arenaria*, or *Teredo*

---

* Since this paper was read, I have found a paper by Dr. Buist on the geology of Bombay at page 167 of the 10th volume of "Transactions of the Bombay Geographical Society." The converse suggestion is here made, viz., that these tubes are the casts of borings made by marine worms through mangrove roots which have been formed by the infiltration of lime held in suspension by rain-water, and deposited in successive layers each monsoon within the outer circumference of the original boring. This theory seems to me as untenable as the other, and formed only for the purpose of accounting for the concentric structure of most at least of the thicker tubes. In the first place, though the fragments of tubes are literally innumerable, I have never yet found one sticking in a piece of wood. In the second place, if the rain-water filtering through the soil of the Byculla Flats were so strongly charged with lime, we should expect to find everything in it thickly coated with lime, not merely the *inner* surfaces of these tubes. On the other hand, the concentric structure of the tube seems capable of explanation by the act of the animal itself in thickening the tube inwards at intervals, for the purpose of reducing the size of the orifice as it diminished in size itself, in the manner pointed out by Sir Everard Home in his paper mentioned below.

† At the time of reading the stick was pulled out, and found to have been roughly cut to a point, thus establishing the truth of the cooly theory.

gigantea, of the habits and history of which very little seems as yet to be known to conchologists.

I am led to this conclusion by the discovery of such specimens as those in the group marked No. 6, showing a *septum* or division running longitudinally down the tube for some distance from the small or upper end in such a manner as to divide the tube into two. These two divisions are in fact the cases of the two siphonal tubes of the animal—one respiratory and the other excretory—which were closed at will by means of two triangular pallets working loosely within the shell. The union of these two tubes into one through the greater part of their length is the characteristic feature of the Pholadidæ. To the family of Pholadidæ the Teredines are now determined to belong. But it was long before the Kuphus, which I believe these specimens to be, was admitted to a place among the Teredines. M. Rang, who under the name of Septaria excluded *Teredo gigantea* (Kuphus) from the genus Teredo, while observing that it very closely approximates the Teredines and the Fistulanæ, thus describes it :— "Animal unknown ; shell unknown ; tube calcareous, thick, solid, in the shape of a very elongated cone, and irregularly flexuous, furnished internally with small incomplete annuliform septa, terminated at one of its extremities by a convexity, and at the other by two slender and separated tubes."

Rumphius figures, under the name of *Teredo arenaria*, a species of tubular shell found in shallow water, among mangrove trees, apparently identical with that described by M. Rang, and represents the double tube at the smaller end as branching into a distinct bifurcation. This of course is a material difference from the specimens now before you.

Lamarck, still excluding this species of Septaria, which he calls Arenaria, from the Teredines, recognised only two species of Teredo, *viz., Navalis* and *Palmulatus*. The latter he thought differed only in its greater size from the former, which is the species long and unfavourably known to sailors as the borer through the bottoms of wooden ships.

In 1797 Mr. Griffiths discovered at Sumatra a tubular shell apparently of a species nearly identical with these before you. He noticed the difference in the structure of the double tube at the smaller end between his specimens and those figured by Rumphius, but ascribed it to the difference of situation in which they were found. Mr. Griffiths' specimens were procured from a small sheltered bay, with a muddy bottom, surrounded by coral reefs, on the island of Battoo, near Sumatra, which was exposed by a violent earthquake. The largest was 5 feet 4 inches in length and 9 inches in circumference at the base, tapering upwards to 2½ inches. Most of them were covered with small Oysters and Serpulæ for about a foot from their upper extremity, showing that they must have protruded that distance from the muddy bottom upwards into the water. But owing either to the depth or the muddiness of the water, they had escaped notice till the natural convulsion which laid bare the bottom of the bay. Mr. Griffiths remarked that the large end was completely closed, and had a rounded appearance and was very thin, while the small end was very brittle and divided by a longitudinal septum running down for 8 or 9 inches. Many of the shells he described as nearly straight, while others were crooked and contorted. The substance of the shell he described as having a fibrous and radiated appearance. And herein lies the only essential difference between his specimens and these before you, which for the most part present a concentric, not radiated, appearance.* In all other particulars they approach very nearly to Mr. Griffiths' Battoo Shells, except in their smaller size.

Godfrey Sellius had been the first in 1733 to recognise a true bivalve mollusc in Teredo. But it was reserved for Sir Everard Home, R. N., in 1806 to discover a species of Teredo in the shells Mr. Griffiths had found at Battoo. He bestowed on it the name of *Teredo gigantea.* He published his discovery in a paper entitled " Observations on the Shell of the Sea-worm found on the Coast of Sumatra, proving it to belong to a species of Teredo," and presented the specimens from Battoo, as well as others found in " Another inlet of the sea, sticking out from rather hard mud mixed with sand and small stones from 8 to 10 inches or more and from 1 to 3 fathoms under water," to the Museum of the Royal College of Surgeons. Unfortunately these have now all disappeared, except two marked E348 and E 349c. They are thus described in the Catalogue :— " E348.—*Teredo* (Furcella) *arenaria*—Rumph. sp. (*Teredo gigantea*, Home). Habitat : Indian Ocean. Presented by Capt. Sir E. Home, R. N. E349.—Specimens marked *a* to *i.* *c.*—The terminal portion of the shell and the double tube."

---

* In some instances the outer and inner layers are shelly and those in the centre distinctly crystalline. In a few the texture of the shell is crystalline throughout. The animal could not have formed a crystalline shell. But by the action of the mud or water in which it was deposited the shell may have been cystallized, as I have pointed out is common with those found near the brick-fields on the west of the Flats; and the crystals may have been subsequently decomposed by some other influence on the surfaces exposed to it.

Among the lost specimens are some "showing the manner in which the animal closes the tube with transverse septa at certain periods of growth" and "the pallets which are attached to the base of the tube."

Henceforth the right of this mud or sand borer to a place in the genus Teredo and family Pholadidæ appears to have been always recognised. How or where he got the name of Kuphus, or Uuphus, or Cyphus, for there seems to be some uncertainty in regard to its spelling, by which he is known to Gutteard, I cannot say. But Sowerby in the "Thesaurus Conchyliorum" thus describes two species, *giganteus* and *clausus*, of a genus 'Kuphus,' Gutteard, synonym, ' Furcella', Oken :—"The tube of this sand-burrowing mollusc attains the length of some feet, and has been known as the gigantic septaria of Lamarck. The small end which protrudes from the surface of the sand is divided by a central septum, and sometimes forms a double separated tube. The pallets of the larger species only are known ; they are spathulate and deltoid. No valves have been found of either species."* The only other known specimens of Kuphus in England beside the two I have just mentioned in the College of Surgeons are those in the British Museum. In the family Pholadidæ, next to genus Teredo is a specimen marked "Cuphus (Gutteard)." Under it is written " *Furcella arenaria* (Gutteard sp.)." It is a piece of tube 15 inches long, with a closed and rounded end showing a visible suture. By it are two pallets. There is nothing to show where any of these specimens was found, and no one at the Museum seemed to know. Besides these, in another show-case are two very long and big pieces of tube, wanting the round ends, but showing well the longitudinal division into two at the smaller end. One of them is labelled "Singapore." The other, 3 feet 9 inches long, has no history ; but Mr. Smith, the Curator, believed both pieces were obtained from Mr. Charlesworth, a well-known Geologist.

Woodward in 1854, writing of Teredo, after describing *T. navalis* and *corniformis*, continues :—"The tube of the Giant Teredo (*T. arenaria*, Rumph., *Furcella*, Lamarck) is often a yard long and 2 inches in its greatest diameter. When broken across, it exhibits a radiating prismatic structure. The siphonal end is divided lengthwise, and sometimes prolonged into two diverging tubes." In 1885, Wm. Clark wrote an account of Teredo, in which he suggested that certain points

---

* I have never found any pallets. If they do not exist, it may be because the Byculla Flat specimens belong to the smaller species, which to judge from the size of those found would appear to be the case.

of analogy of Teredo and Dentalium make it appear that the former is the passage between Lamellibranches and Gasteropods ; that is to say, putting the proposition in a more popular form, Kuphus may be regarded as the connecting link between bivalves and univalves. Lastly, in May 1875 was published an illustrated paper on Kuphus in Reeves' "Conchologia Iconica" (probably written by Mr. Sowerby, Mr. Reeves having died in 1865), which thus describes the genus Kuphus, Gutteard Cyphus :—" Mollusc ; sand-burrowing ; tube large, white, rough, slightly ringed; posteriorly attenuated ; divided interiorly into two tubes ; chambered transversely with septiform laminæ ; valves unknown; compressor palmets shelly deltoid."

The writer goes on to point out that the general appearance of the tube is so like that of the Teredo as to leave little doubt of the nature of the animal and its affinity with the genus Teredo. At the same time he says it can hardly be included in that genus, the valves having never been seen, and it being certain that the animal does not bore like the Teredines.

He figures two species, *gigantea* and *clausa,* the former of which has the lower end broadly open, the latter closed in a rounded oval disc with a visible suture.

It is hard to see how, with a closed and rounded end, *inside* which the valves, if any, must be situated, the animal can have conducted its boring operations through the mud, especially as the shell at this part is described as very thin. I have a theory, of course a mere guess, as I have never seen the anterior extremity of the shell, which you may think it presumptuous in me to advance, but still it does seem to me not impossible that the closed end may be not the characteristic of a species, but due to the act of the individual. We have noticed the rounded projections, the shell of which is very thin, occurring wherever the animal stopped progressing in the old direction and started in a new. Suppose for any reason he did not start again, the tube would end in a rounded projection. Might not this account for the rounded ends of some specimens? Sir Everard Home, in his paper already mentioned, says that both *Teredo gigantea* and *Teredo navalis,* when arrived at their full growth, close up the ends of their shells, and that death is not the consequence of this act. In some of Mr. Griffiths' specimens he says the shell was considerably thickened at the end, and in a few the animal had receded up the tube, forming new inclosures more than once, and at the same time thickening the walls of the tube so as to diminish the canal in

proportion to the diminution of its own size, showing that it must have
survived for a considerable time the first closing of its lower
end.

You thus see how little is as yet known of this last discovered species
of the genus Teredo. But this scantiness of information and paucity
of specimens may be attributed rather to the inaccessibility than to the
numerical scarcity of the creature. For an animal that bores several
feet deep into a muddy bottom several feet below water cannot be
said to offer much encouragement, at least to human beings, to make
his acquaintance. But the introduction may in a measure be facilitated
if the tubes to be found in such numbers about the Byculla Flats are
in fact the remains of this creature. That they are, I think, may
be inferred. First, from the similarity of the place in which they
are found to that described as the home of Mr. Griffiths' "Sea-worm."
At the time when the Indian Ocean ebbed and flowed across the
Byculla Flats, their condition must have nearly resembled that of the
shallow sheltered bay, with a muddy bottom, in the neighbourhood of
Sumatra. Secondly, the general appearance of the shelly tubes here
agrees with the descriptions I have quoted to you in every point,
except that the structure is concentric instead of radiating, which
may be due either to a difference of species or to the alteration the
shells have undergone in the process of fossilization.* Thirdly, and
most important, we recognise here the longitudinal septum, dividing
the tube into two for some inches of its length, which characterized the
shells discovered by Mr. Griffiths.

It is true that in the descriptions and specimens I have mentioned
of Kuphus, there are the closed and rounded lower ends which I have
not yet succeeded in finding. But these are probably still awaiting
discovery some few feet lower down. Major Frere tells me he found
one, but I am sorry I never saw it. I have found these two specimens, marked No. 7, which at first I was inclined to hope might be
the extreme tips of the rounded ends, the shell of which you will
remember is described as being very thin. I am however now inclined
to think that they are nothing more than the excrescences, which we
saw the animal threw out in his shell whenever he changed his direction, and which have been knocked off the tube. The group of

* If the process of crystallization were gradual, and the crystal were substituted for the
shell in successive layers, but were afterwards decomposed, say by heat, the structure of
the tube would be concentric and the texture non-crystalline. If the heat were not sufficient to penetrate the whole thickness of tube, the centre layers would still be crystalline, as first altered from the shell, and the outer and inner non-crystalline, not according to the original structure of the shell, but owing to the second alteration it had undergone from its crystalline shape.

specimens, marked No. 6, shows how prominent some of these excrescences are, and how easily in consequence they might be knocked off.

I think, from the very large number of the tubes now to be found loose on the Byculla Flats, that the animals inhabiting those tubes actually lived in the soil of which the Flats are composed, and that if we dug further down we should come upon their lower ends. But of course it is possible that the fragments of tube now found on the Flats were only washed in from outside, and that the lower ends are still sticking, head downwards, in some other soil. The fact that we have only found upper ends as yet on the Flats lends colour to this theory. The two specimens, marked No. 9, however still show the sort of soil in which to look for the animal. These are evidently lumps of clay, though now considerably indurated, through which, while soft, the creatures which inhabited these tubes were working their way.

Bearing this in mind, and remembering the description of their habitat as given by Mr. Griffiths, I would recommend careful search to be made in those sheltered bays and creeks, which so abound in the neighbourhood of Bombay, with muddy bottoms over which the tide flows with no great violence to a height of from 6 to 15 feet. If once we can find their home, there will be no lack of specimens, for the abundance of remains on the Byculla Flats shows that, in numbers at least, they are not deficient. A large capture of living specimens would probably be attended with important results to science generally. It would certainly be a valuable addition to the best collections in England, and would reflect great credit on this Society. It will however be necessary to remember that as all that glitters is not gold, so every tube is not a Kuphus. Here is a small group of specimens, marked No. 10, which are the tubes of Dentalia, also very common on the Byculla Flats. One you see, comparatively modern, is hardly altered at all ; but the others from their appearance might be coeval with the oldest and most altered of the specimens of Kuphus to which I have introduced you. The Dentalia, you will remember, are the creatures referred to by Clark in propounding his theory that " Teredo is the passage between Lamellibranchs and Gasteropods." They also are very interesting creatures, because, if Gasteropod at all, they are very exceptional members of that order· Huxley regards them as Pteropods. They constitute a very lowly-organised group without distinct gills or heart and with a but imperfectly developed head. The slender tubular shell, as you see from these specimens, is curved, tapers suddenly, has no division, and has an aperture at each end, that at the smaller being quadrangular, features which readily distinguish it from Kuphus.

But I must not allow myself to be betrayed into trespassing further on your patience. If I were to attempt to describe all the shells to be found on the Byculla Flats I should never have done. Among them I have no doubt are many besides Kuphus that have hitherto enjoyed a reputation for scarcity, simply because by their inaccessibility they have been seldom seen and little studied. By conveniently investigating these remains at our leisure on dry land, we may learn so much of the history and habits of the animal as to be able more readily to secure living specimens in the neighbourhood.

J. B. H.

24th July.

THE BIRDS OF SOUTH GUJERAT.
By H. LITTLEDALE, BARODA.

In Major E. A. Butler's excellent list of the "Birds of Sind, Cutch, Kathiawar, North Gujerat, and Mount Abu" (in the *Bombay Gazetteer*) several birds are omitted which have been found in South Gujerat and the Panch Mahals, and which I think must certainly extend to North Gujerat and the Rajputana Forests at least, if not to Eastern Kathiawar also. The fauna of any district will obviously be intermediate between the faunas surrounding it, and one cannot draw a hard-and-fast line beyond which birds are *never* found to travel. In fact "never" is a word that the Ornithologist should specially beware of; with birds "the world is all before them where to choose," and they exert their privilege of choice to an extent that often upsets the dogmatic Naturalist, whose "never" has to be modified into "hardly ever" to suit the facts of the case.

12. *Falco babylonicus* (Gurney).—The Red-cap Falcon is only recorded by Major Butler from Sind; but Mr. Doig shot one at Sanand, near Ahmedabad.

27. *Aquila mogilnik* (S. S. Gm.).—Mr. Doig has shot at the same place; Butler only records it from Sind.

35. *Limnœtus cirrhatus* (Gm.).—The Crested Hawk-eagle Major Butler records from "Mount Abu, rare." Mr. Davidson writes to me that "it must breed with you; it is the common Eagle in West

Khandesh, and from our hills, Pavagarh (a mountain 28 miles north-east of Baroda), is seen;" but I have not yet found it. It probably will turn up in the hilly forests of Chota Udepur and the Panch Mahals.

39. *Spilornis cheela* (Lath.).—The Crested Serpent-eagle Butler records only from "Sind, rare." Mr. Barnes (*Birds of Bombay*) says it "is very rare; one was obtained at Savantvadi by Mr. Crawford; and another in Sind by Mr. Blandford: these are, I believe, the only recorded instances of its occurrence within our limits." I shot a female and got an egg in a nest at Pattra, 15 miles from Dohad, Panch Mahals, 12th April 1886. Mr. Doig and I were both of opinion when examining it in the flesh that this bird was true *cheela* and not *minor*, and so I think its right place is in the museum of our Society, where it will be found by any one wishing to verify the record, which, as we had only measurements to go by (Hume, *Rough Notes*, Jerdon and Barnes being consulted), and no skins to compare, would be desirable. The nest was in a fork of a *Kodai* tree, in thin jungle, 20 yards in from the flank of the bed of the Anas River. It was a poor straggly affair, not bigger than a Kite's, and hardly so compact. The egg, handsomely blotched and streaked with dark red at the larger end, measures 2·6 × 2·2. On the 25th of May I saw a pair of either this species or *S. minor* feeding a young bird near Beecheewara (Dungarpur, Meywar).

57. *Pernis ptilorhynchus* (Tem.).—Major Butler records the Crested Honey-buzzard from "Mount Abu, rare." Mr. Doig tells me that he has shot it in the Ahmedabad District; he and I found a nest, one egg, in a high *Kadai* tree in thickish jungle at Singargarh, near Saonth, Panch Mahals, and shot the female on the 25th April 1886; and we saw another at Saran, near Dungarpur, Meywar, 5th May 1886. The egg was white, faintly marked with cold brown at the larger end.

65. *Syrnium ocellatum* (Less.) is said by Butler to be a "permanent resident (I believe)." I found its nest, two eggs, 4th March 1886, near the Race-course, Baroda. One egg was much harder set than the other, and had a bloodstain on it from the remains of a half-eaten squirrel that lay beside it. In 1885 I was too late for this nest, finding one fluffy little fellow snapping his bill at me when I called on the family on the 31st March.

72. *Ketupa ceylonensis* (Gmel.).—"Sind, rare," says Major Butler; "has not yet been recorded from Gujerat, neither did I meet with it in Rajputana or Central India" says Mr. Barnes. Mr. Doig and I saw three, and shot one adult and one young bird at Saran, near

Dungarpur, Meywar, 7th May 1886 ; and Mr. Doig shot a specimen
at Harsole, near Ahmedabad, in 1884. The young one at Saran
seemed about four months' old.

74. *Scops pennatus* (Hodgs.)—" Sind, cold-weather visitant, rare,"
is all Major Butler records; and Mr. Barnes says it " occurs sparingly throughout the district, except perhaps Gujerat." I therefore
record that on the 8th February 1886 I shot one, in the rufous
phase of plumage, at Pavagarh, on the hill-side above Champanir,
and my shikarry said he saw another which was *white* (*i.e.*, the adult
phase).

75*ter*. *Scops bakkamuna* (Forst.)—Mr. Doig got a family of six of
these at Saran, and I kept one of the young ones alive for several days ;
they are only recorded from Sind and Abu, and with nocturnal species
every occurrence is worth record. The nest-hole was in a high *Mowra*
tree, and was inhabited also by a colony of tree-ants, who made it
uncommonly hot for the man who got down the Owlets for us ; in fact
he twice " resigned," but the sight of a depreciated " dib" encouraged
him to persevere and succeed at last.

77. *Glaucidium radiatum* (Tickell).—Butler only records this from
the jungles at the foot of Mount Abu ; but we found it common in
the mahals from Dohad northward to Saran (Meywar); and *A. brama*
correspondingly scarce, and only near the villages.

98. *Cypselus melba* (Lin.).—I only mention to protest against
Major Butler's remark "only occurs, as a rule, in Gujerat, *within reach
of the hills.*" As the Gujerat Alpine Swifts are within reach of the Himalayas if they choose to go there to roost and return in the morning,
this seems an unnecessary limit to place on the range of birds with
such wonderful powers of flight ! I have frequently seen them overhead near Baroda, and have shot them on the 21st September.

104. *Dendrochelidon coronata* (Tick.).—This lovely bird is not
in Butler's list ; but it is quite common in the hill jungles of the Panch
Mahals, especially near the tanks in those jungles. I found a nest with
egg on a thin bough of a leaf less tree, 20 feet above the path in
the midst of jungle, near Saran. The nest was hardly 1½ inch in
diameter, including the bough to which it was glued ; and both nest
and egg are safe and sound in my collection—a feat which Mr. Hume
(*Nests and Eggs*) never managed to accomplish, and he says " it
is almost impossible to get the egg (for they lay only one) down
unbroken.

118. *Merops philippinus* (Lin.).—Major Butler only records from

" Mount Abu, rare, occurring only as a straggler." This leads me to remark that Major Butler does not appear to have fully worked out these species, that keep along rivers such as are more common in South than in North Gujerat. This species is common enough, and breeds along the Mahi from the mouth nearly to the source ; it has to keep to the larger rivers during the breeding season (May), leaving them for the meadows during the rains.

124. *Coracias garrula* (Lin.).—Butler says " Sind ; seasonal visitant ; not common." Mr. Doig notes in my copy of Barnes : " very common in Gujerat, the Ahmedabad districts, in August and September, and again in February ;" and I saw two at Goblej, near Khaira, September 27th, 1886.

127. *Pelargopsis gurial* (Pears.).—The Stork-billed Kingfisher is not recorded by Butler ; but we found it along the Mahi in the Panch Mahals (and see my paper in No. II. of this Journal).

147. *Palaeornis eupatria* (Lin.).—Butler refers to one Sind specimen of doubtful authority. Mr. Murray (*in Epist.*) says " this was undoubtedly a cage-bird escaped ; tail feathers much abraded."

· 164. *Yungipicus nanus* (Vig.).—" Mount Abu, rare," says Major Butler. Mr. Doig saw a pair, and shot a male near Ganji, Dungarpur, Meywar, 4th May 1886. It measured only 4¾ inches in length.

193*bis. Megalæma inornata* (Wald.).—Common in the jungles of the Panch Mahals and at Pavagarh.

238. *Dicaeum minimum* (Tick.).

240. *Piprisoma agile* (Tick.).—Neither of these little flower-peckers is in Major Butler's list ; they are both permanent residents about Baroda.

250. *Sitta castaneiventris* (Frankl.).—Not in Major Butler's list. I shot a pair at Saran, Dungarpur, Meywar, and saw two others there 5th May 1886. They did not appear to be breeding then.

· 268. *Volvocivora sykesii* (Strickl.).—Major Butler records from " Abu and the low hills east of Deesa ; rare." It goes east after the rains, and I saw it not unfrequently in the Panch Mahals in May, doubtless on its way west to breed, which it does about Baroda.

· 285. *Dissemurus paradiseus* (Lin.).—Not in Major Butler's list, but " breeds in the east of Godhra, and therefore probably throughout the Panch Mahals" (J. Davidson, Esq., c.s., *in Epist.*) ·

293. *Leucocerca leucogaster* (Cuv.), which Major Butler only records from Abu, breeds at Baroda also, through rarely, *L. aureola* being by far the commoner species.

297.   *Alseonax latirostris* (Raffles.).—Is not in Major Butler's list;
and Mr. Barnes says " it has not been recorded from either Sind or
Gujerat." I found it common at Saran in Meywar ; shot a female,
May 9th, 1886.   It is so like a Sparrow that doubtless it has been
often overlooked, and will probably be found in quiet shady places,
over water, throughout the jungles of the Presidency, except Sind.
Although in appearance like a Sparrow, its manners resemble those of
305, *Cyornis tickelli*, especially in its robin-like flutter of the wings
when standing.   I saw it *whacking* some insect several times on
a bough, just as a Wood-shrike does, and then swallowing the big
morsel whole.

452.   *Ixos luteolus* (Less.).—Not in Major Butler's list, but
common about Baroda and in wooded ravines throughout the
district.

459.   *Otocompsa leucotis* (Gould).—Though common in the more
desert tracts to the north, I have never seen this bird in the park-
like country south of Ahmedabad to the Nerbudda.   Mr. Barnes
however says it " is far from being uncommon in Gujerat."

463.   *Phyllornis jerdoni* (Blyth).—Not in Major Butler's list,
but nevertheless occurring sparingly about Baroda, and more commonly in the forests of the Panch Mahals.

467 & 468.   *Iora zeylanica* (Gmel.), which Major Butler records
only from Abu, and " not very common." there, is very plentiful
about Baroda, where I have found many of its nests.

475.   *Cospychus saularis* (Lin.).—Major Butler calls this a " cold-
weather visitant."   A pair have just left my porch with their
young family which they reared there this June !   I saw seven adult
birds together in a mango grove at Jhalod, Panch Mahals, 20th
May 1886.

481.   *Pratincola caprata* (Lin.).—Major Butler calls this a permanent resident in Gujerat ; but it certainly is not found in the Baroda
District from April to September, and though we specially watched
for it, neither Mr. Doig nor I saw *one* in the Panch Mahals last
April and May.

490*ter*.   *Saxicola capistrata*. }   Not in Butler's list, but recorded from
517.   *Lusciniola neglectus*. } Sind in Murray's *Verteb. Zool. of Sind.*

553.   *Hypolais rama* (Sykes).—Although the *Phylloscopinae* are all
cold-weather visitants, it is very probable that others breed in Sind
besides this species, which Mr. Doig found breeding plentifully
there.

558. *Phylloscopus lugubris* (Blyth).—Not in Butler's list, but I shot one out of a flock of five near Baroda Race-course, 17th September 1885; and Mr. Barnes says " very rare winter visitant to the Deccan," which give us two landmarks on its line of migration.

582*bis*. *Sylvia minuscula*⎫ Neither in Butler's list; both in
582. *S. althæa*⎭ Murray's *Verteb. Zool. of Sind.*

560. *Phylloscopus viridanus* (Blyth).—Not in Butler's list.. Shot one at Pattra, near Dohad, Panch Mahals, 14th April 1886. Mercly a cold-weather visitant to the Deccan " (Barnes).

631. *See* previous paper, Journal No. II.

647. *Machlolophus xanthogenys* (Vigors).—Not in Butler's list. Mr. Doig shot a male in a mango grove at Jhalod, Panch Mahals, 21st April 1886, evidently breeding or about to breed; and we saw a pair at that " bird paradise" Saran,* near Dungarpur, about ten days later.

674. *Dendrocitta rufa* (Lath.).—To my previous paper (Journal No. II.) let me add regarding this bird that I counted twenty-three (23) of them fly out of one tree at Kadana on the banks of the River Mahi, Panch Mahals, 28th April 1886, and found them very common in the jungles between Dohad and Khairwarra at that time.

Serinus pectoralis (Murray), *sp. nov.*—Not in Butler's list. (*See Verteb. Zool. of Sind*, 193, as also p. 201 for 784, *Palumbus casiotis.*)

765. *Spizalauda deva* (Sykes).—Not in Butler's list ; but this is the commoner sort about Baroda, and *S. malabarica* the rarer.

805 & 306. *Cyornis tickelli* (Blyth).—Common in secluded spots, near water, throughout Gujerat, though not recorded from that district by Major Butler.

839. *Sypheotides aurita* (Lath·).—Have found it breeding about Baroda at the following dates :— 19th August 1885.—Two ˈeggs, and a third, a bright green colour, extracted 21st September 1885.—

---

* There is a stream from a spring here, with overhanging trees, and not another drop of water for miles around. The little stream is only about 5 yards broad, and after a course of 300 yards or so disappears in the sand; but I noted in my diary at the time the following birds in that one little oasis :—Green Barbets (*inornata*), Coppersmiths, Common and White-bellied Drongos, Dover, Green Pigeons, Nuthatches (250), Tickell's Blue Redbreast, Titmice (Grey and Yellow-cheeked), Orioles, Koels, Crows, Sparrow-hawk (on nest, three eggs), Owlets (*A. radiata, Scops bakhamuna*), Paradise and Fantail Flycatchers, Kingfishers (*P. gurial, C. rudis, H. smyrnensis*, and *A. bengalensis*), Woodpeckers (*Aurantius* and *Mahrattensis*), Common Sand-pipers, Lapwings, Painted Sand-grouse, common Mynahs, White-throated, Hodgson's and Stewart's Wren-warblers, Tree-pies, Common and Yellow-throated Sparrows, Bulbuls (462), Brahminy Mynahs, Fish owls (*K. ceylonensis*), Crested Tree-swifts, Crested Honey buzzards, Babblers, seven large Grey˙Cuckoo-shrikes, Magpie Robins, Green Bittern, Rose-ringed and Rose-headed Parrakeets, Mottled Wood owl, Indian Nightjar—what a choir !

Caught three chicks just out of shell; no nest; fragments of shell on
a flat bit of ground amid thin grass. 9th August 1886.—Four fresh
eggs. 13th September 1886.—Three fresh eggs. All the foregoing
from near Bakrol, six miles from Baroda.

842. *Glareola orientalis.*
845bis. *Ch. pluvialis.* } Not in Butler, *vide* Murray, *op. cit.*
847. *H. ventralis.*

843. *Glareola lactea* (Tem.).—The Lesser Swallow Plover not in
Butler's list, though common in the sandy, rocky bed of the Mahi above
Wasad. I got 18 eggs in the bed of the Mahi above Sihora, 6th April
1886. There were no nests, and the eggs were either single or
in pairs on islands. Some were far in under the ledges of rock;
others right out on the gravel; and the sheltered eggs were far
finer coloured than the exposed ones.

900 *Parra indica* (Lath.).—Butler says "permanent resident,
I believe." It breeds commonly about Baroda, laying its eggs on
the floating lotus leaves. People in India generally call this bird
a Jacana, pronounced Jakana; but the name is spelt Jaçana in Coues's
Birds of North America, and that indicates the correct pronunciation
I believe, though Ogilvy's dictionary pronounces it as *Jakana.*

924bis. Not in Butler, *vide* Murray, p. 270.

932. *Ardetta flavicollis* (Lath.).—Only recorded from Sind by
Major Butler; but Mr. Doig got it near Ahmedabad; and I saw
a pair near Baroda, May 1884, but as I was waiting for a Panther
(that never came), I did not secure a specimen.

850. *Ægialitis minuta* (Palls.).—To the instances recorded by
Major Butler I may add that I have frequently shot it along the Mahi,
and found two nests, three eggs each, last April 6th, at Sihora. On
the Mahi south of Dakore on the same day I found three nests, two
eggs each, of *Æsacus recurvirostris* along the river-bed, thus
justifying Butler's remark of this species, (858) "permanent
resident, *I believe.*"

NOTE ON A RECENT PAPER BY DR. BONAVIA ON THE MANGO.

By Surgeon K. R. Kirtikar, I. M. D.,

Acting Professor of Anatomy, Grant Medical College, Bombay.

Under the presumption that he was presented with real Bombay
mangoes. Dr. Bonavia without reserve declares that they were

disappointing. He describes them as having a red cheek and yellow colour ; they were stringy. The very fact of their being stringy precludes them from being considered the real Alphonso mangoes, much less could they be considered the best. Any mango grown in Bombay, or around Bombay, may have a red cheek and yellow colour ; but that does not make it a good mango. The entire absence of strings is *the* characteristic of the real Bombay *Alphonso*, or *Afoos* as it is popularly called. The mesocarp, or rather the sarcocarp, consisting of the pulp of the fruit, can be cut through like fresh cheese that is not very hard, or can be easily scooped out by means of an ordinary dessert spoon with a clean cut. As regards the real mango being inferior in flavour to the scores of varieties Dr. Bonavia has seen in Upper India, even supposing he has tasted the best Bombay mango, it is a mere matter of taste. There is no accounting for tastes. There is room for wide varieties. The common Konkani kunbi will never care to eat the finest table-rice that a high class Hindu would prefer. The kunbi would prefer his coarse rice, which he declares is sweeter and more substantial. Children will never eat, at any rate fully appreciate, the real *Afoos*, but will be content to suck the juice of the *Raiwal* or smaller varieties of mangoes. The real Bombay mango is luscious, sweet as honey, and its epicarp or rind very thin, almost transparent. The thinner it is the better, and such as can be easily peeled off without tearing through the rich and succulent pulp. It does not matter then whether it has a red cheek or not, or whether it is yellow, or rich orange, or saffron coloured. To turn out a good mango, free from acidity, the mango must be plucked at the proper time. The nearer it is to the ripe condition while yet on the tree the better will it turn out. If the mango is plucked immature, even if it be if the best kind, it will fail to give satisfaction. It will often, near the stone and a portion of its pulp, remain pale in appearance, and often form fibrous cavities, and will be acid to taste, showing that there has been a localized gangrene of the parts concerned. A good mango on the other hand, plucked perfectly mature and about to ripen, will require certainly not more than five, six, or seven days at the outside to be fit for the table. " The mango may bear," I agree with Dr. Bonavia, " being plucked under-ripe, and can *easily*"—so far as transit is concerned I think—" be sent to England and there ripened," but I question if it would ever ripen under such circumstances to perfection. A good mango can never ripen well, much less to perfection, under the chilling influence of the cold used to preserve it. Cold may prevent decay and decomposition, but I doubt whether will ever hasten ripening or help it. I think it deteriorates the fruit. Cape pears may find a market in Covent Garden, and so would Bombay mangoes with a brisk journey of nineteen days across the continent if carefully packed and looked after constantly during the Red Sea voyage and continental journey. But in my opinion there would always be a difference between a fresh mature mango ripening under natural processes in five days and an under-ripe mango ripening in twenty days under forced conditions and chilling preservative influences.

There is often so much deceit practised by the mango-sellers in the bazaar that an unwary and uninitiated foreigner is likely to be taken-in and presented with any wretched mangoes—perhaps some thick skinned Goa mangoes—under the name of Alphonso mangoes. But anybody that knows what a real mango is, from its taste, appearance, flavour or aroma and texture, will always recognize it. Even the feel is characteristic ; and the smell, without cutting, is diagnostic. The first gatherings of these mangoes are always defective and sold at enormous prices, and Dr. Bonavia has a just reason to complain when he finds that Rs. 6 have to be paid for a dozen mangoes. People are so impatient to eat the first fruit of the season that they pay down any price. The agents of the up-country Rajahs buy them up at fabulous rates, as the Rajah's money is almost without a guardian in such cases· Induced by the hope of making an easy fortune, the mango contractor takes the earliest opportunity to have his pick of the fruit, and in doing so often plucks under-ripe mangoes, which sometimes never ripen at all or, if they do, do not develope into the perfect fruit and are insipid. Sometimes they rot during the ripening process. I have had an opportunity of tasting some Upper India and Sind mangoes, and the Deccan, Goa, and Bangalore ones are common enough, but they do not come up to the Bombay fruit. It is not my intention at present to write anything on the different varieties of the mangoes found in Bombay. During the next mango season the Bombay Natural History will hold an exhibition of the different varieties of the mango, when it is hoped a careful list of the various kinds will be made out.

I come now to another part of Dr. Bonavia's remarks. When he says that he has preached for many years that " it is a grave mistake to throw away the thousands of stones of superb mangoes that are consumed every year," one would think that Dr. Bonavia has practical experience in the matter sufficiently strong to substantiate his remarks. He is clearly mentioning what is contrary to the actual experience of mango-culturists on this side of India when he says that "it does not at all follow that a stone of a good mango will not give a *better* fruit than that of its parent." The common experience here is that a seedling is not only not better than its parent in the production of the proper fruit, but as a general rule is not even as good as its parent. For instance, a good Alphonso or Pâyari (spiked or sharply curved at the apex) can never be cultivated out of its respective seedlings. They always degenerate, no matter what the parent is. A special mango has always to be obtained from grafts. Grafting mangoes is an industry which is very paying, and now that the whole island of Salsette is under extensive cultivation at the hands of intelligent and painstaking landowners, it is certain that at no distant day Bombay will be abundantly supplied with excellent graft Alphonso and Pâyari mangoes. Notwithstanding the high authority of DeCandolle, quoted by Dr. Bonavia, with regard to the mango cultivated in the colony of Cayenne bearing stones *which produce better fruit* than that of the original stock, the common experience in India with regard to the Alphonso or Bombay mango is different. The seed as it developes into a plant takes a long time to bear fruit, the fruit itself losing the characters of its parent The seed of an Alphonso mango will not produce an Alphonso fruit, but degenerate into a common Raiwal.

K. R. K.

A CATALOGUE OF THE FLORA OF MATHERAN.

BY THE HON. H. M. BIRDWOOD, VICE-PRESIDENT.

A RECENT visitor to Matheran is said to have complained sadly of the monotony of its vegetation. That too familiar "Matheran tree" was everywhere, and everywhere the same ; and though it was very beautiful, with its glossy leaves and purple plums, it so impressed its sameness on the landscape as to induce a sense of depression, from which the visitor was glad to escape. It is just possible that his experience was not altogether singular ; for we do not all cultivate alike the faculty of observation. Two men, with the same love for the beauties of Nature, and with equally good eyesight, may look on the same fair scene of hill and forest, sea and sky, with very different apprehension of its infinite variety, and with very different degrees, therefore, of satisfaction. The one may take in, with the trained eye of the artist, notable details which the other misses. He will see

wondrous shapes and colours, and gradations of colour, in every wave
and cloud, and leaf and boulder, where the other sees only trees on the
steep hill-side and a waste of water dappled with shadows.   It is one
of the main advantages of our Society that it teaches its members to
make a right use of their eyes; and in some of us, the discovery that
even blades of grass are not all alike may perhaps have awakened into
activity a faculty hitherto dormant.   So that now, in our continued
researches in the vegetable world, we become aware of a multitude of
beautiful forms, hitherto unnoticed, which daily reveal themselves
to us; and it is no more possible for us now to be oppressed by the
sameness of our surroundings, whether at Matheran or elsewhere.
But though a thirst for knowledge has been thus created, we cannot so
easily quench it.   We have no leisure for systematic study; and when
we consult our standard authors for information about plants, we are
repelled by a difficulty which meets us at the outset.   We cannot
refer to the works of Hooker or Roxburgh, Brandis, Graham or
Dalzell, with any readiness or comfort, if we have first to find out
laboriously for ourselves the scientific names of plants by which alone
they are generally known to these writers.   Though this difficulty
may be reduced, it is not quite removed by the use of such a synopsis
of Orders as that contained in the "Artificial Key" to Orders I. to
LXXI. of Dalzell and Gibson, published in 1875 by Captain H. H.
Lee, R. E., or in the Revd. Dr. Fairbank's "Key to the Natural
Orders of the Plants of the Bombay Presidency," published in 1876;
and members of our botanical section are still unprovided with
correct lists of the local names of plants, with the aid of which they
would find it a comparatively simple matter to acquire the information they are in search of.   No doubt, we find valuable
glossaries of vernacular names in Roxburgh and Brandis; but the
names are not always those in use in this Presidency, and the
glossaries do not, therefore, sufficiently meet the requirements of
students of the rich flora of Bombay and its neighbourhood.   And
this remark applies also to the very full list of Bombay names in the
index to Sir George Birdwood's "Vegetable Products," which is
meant for the use rather of the physician, the merchant, and the
agriculturist than of the mere botanist.   It is in the hope, then,
of removing this initial difficulty, to some extent, as regards
the vegetation of a certain limited area, which is much visited
by members of our Society, that I have compiled this catalogue,
which furnishes a ready method of learning the scientific name

of a plant of which the vernacular name is known. Almost every coolie at Matheran knows the names of most of the Matheran trees. Indeed for some plants you may get a brace of names or more, if you will only question your informant long enough. My own particular coolie, Krishna, in the course of two hours spent in the Primeval Forest and below Chowk Point, gave me no less than 75 names, which he told me he had learnt in the forest, with an air as if the trees themselves had told them to him. With full confidence in the sources of his information, I have included these names in the third column of the catatogue and in the index appended to it, with many others furnished by Mr. Jaykrishna Indraji, Curator of Forests in the Porbandar State, a keen botanist, who lent much efficient aid to the late Dr. Sakharam Arjun in the collection of his Bombay herbarium. I am much indebted indeed to him, and also to Dr. Kirtikar, for carefully revising the whole of the catalogue, which can now, with the aid of Krishna or any other hill florist, be used for the purpose for which it is intended. I would only add that those who so use it must not expect to find it by any means a complete list of the flora of Matheran. It is a fair-weather catalogue, written in the month of May and the early days of June, when many plants which blossom in the rains or the cold-weather are dried up, past all recognition. It is a completer list, therefore, of trees and perennial shrubs and climbers than of herbaceous plants; though it contains also the names of a few such plants, inserted either from memory of past cold-weather visits to the Hill, or obtained from friends or from Mr. Campbell's Gazeteer, or the Revd. Mr. Gell's Catalogue, published now many years ago, and afterwards republished by Dr. Theodore Cooke. Such as it is, it is as complete as it could be made in the course of several very pleasant rambles in the company of our Vice-President, Dr. D. MacDonald, Mr. Chester MacNaghten, and Mr. Jaykrishna. Such as it is, I offer it to the Society as an instalment only of a work which I hope will be taken up, continued, and enlarged by others, if not by myself, till we are in possession of tolerably complete catalogues of the flora of all parts of the Presidency. I can only hope that members of the Society who have the good fortune to visit Matheran during the next six months will remorselessly criticize and amplify my work and favour our editor in due course with the result of their labours. To this end, I have asked Mr. Sterndale to issue a few interleaved copies of the catalogue in pamphlet form, and these can be procured from the Secretary.

CATALOGUE.

NOTE.—In the first two columns, the nomenclature adopted for the first 51 orders, exclusive of Order 33, " Loranthaceæ," is that of Hooker's " Flora of British India," Vols. I—IV, which do not include " Loranthaceæ," or the Orders 52—78 represented in this catalogue. The synonyms given in the second column are the names under which the plants are described in Dalzell and Gibson's " Bombay, Flora." Where no synonyms are given, the plants are described under the same names in both Hooker and Dalzell. In the third column, the names are spelt, for the most part, according to the Hunterian system. The word " vel" or " yel," which recure frequently as a component part of a name, means a " creeper " or ' climber.' The words " lahan" and "dhakta" (fem. " dhakti") mean ' small,' "mota" (fem. " moti") means 'big,' "pandhra" means ' white,' " kala," 'black' " tamra," 'red,' and " karu ' bitter.' The prefix' " ran" indicates a " jungle plant," or as we should say " a wild plant," though all the plants in the list are of course wild or indigenous plants on the hill, with the exception perhaps of the Jack-tree (Artocarpus integrifolia).

Natural Order.	Genus and Species.	Vernacular Name.
1 Ranunculaceæ......	Clematis triloba	Mor-vel, Ránjái.
2 Dilleniaceæ.........	Dillenia pentagyna	Karambel, Dákhta Karmal.
3 Anonaceæ	Uvaria Naram......................	Naram-panal.
„ „	„ lurida	
„	Bocagea Dalzellii .... } Syn. Sagerœa laurina...... }	Sajeri, Hár-kinjal.
4 Menispermaceæ ...	Cocculus macrocarpus	Vátoli, Vát-yel.
„	„ villosus	Tán, Vásanvel (Sanskrit Vásadani).
„	Cyclea peltata	Pár-yel.
5 Capparideæ.........	Capparis pedunculosa	Kolisna.
„	„ horrida	(Near Alexander Point.)
6 Tamariscineæ	Tamarix ericoides	Jao, Sarub, Saráta.
7 Guttiferæ	Garcinia indica	Kokam, Rátamba (Wild Mangosteen).
......	„ ovalifolia .... } Syn. Xanthochymus ovali- folius .... }	Haldi.
......	Ochrocarpus longifolius ... } Syn. Calysaccion longito- lium .... }	Harkia, Surangi.
8 Dipterocarpeæ......	Ancistrocladus Heyneanus ...	Kardor, Kardori. .
9 Malvaceæ............	Hibiscus hirtus	
„	Thespesia Lampas...............	Rán-bhendi, Láhán-bhendi
„	Bombax malabaricum ...... } Syn. Salmalia malabarica. }	Sáwar, Támri sáwar (Silk-cotton tree). (Sanskrit. Rakht-shálmali).
10 Sterculiaceæ	Sterculia guttata	Goldor, Gordar, Kukar.
„	„ colorata	Bhaikui, Khavas, Khaushi.
„	„ urens	Sáldhawal, Karai, Kuari.
11 Tiliaceæ	Grewia tiliæfolia	Dhámal.
„	Erinocarpus Nimmoanus	Chaurá, Chor, Cher.
„	Triumfetta pilosa	Kutre vandre (" Dogs and Monkeys").
. „	„ rhomboidea	Necharda.

Natural Order.	Genus and Species.	Vernacular Name.
12 Geraniaceæ	Impatiens acaulis	Lábán Tírda, Berki.
„	„ oppositifolia.........	Sanmukh-patri, Tírda (Wild Balsam).
13 Rutaceæ	Atalantia monophylla	Mákar-limbu (" Monkey Lime").
	Murraya Kœnigii ....⎫ Syn. Bergera Kœnigii ...⎭	Karepát, Karu-nimb.
	„ exotica (var. paniculata)	Pándri, Kunti.
14 Meliaceæ	Soymida febrifuga..............	Polá·á.
	Chloroxylon Swietenia	Billu, Haldá.
15 Celastrineæ.........	Gymnosporia montana ......⎫ Syn. Celastrus montana...⎭	Yεkdi.
......	Gymnosporia Rothiana......⎫ Syn. Celastrus Rothiana ...⎭	Moti Yεkdi.
„	Hippocratea Grabami	Yεvti.
16 Rhamneæ	Ventilago madraspatana	Kánvel, Lokhandi.
„	Zizyphus xylopyra.............	Guti, Ghuti (Hart Point and elsewhere).
„	„ rugosa	Toran.
17 Ampelideæ	Vitis discolor . ...⎫ Syn. Cissus discolor.....⎭	Telitsa yel.
.. ..	„ latifolia ....⎫ Syn. Cissus latifolia ....⎭	Nadena.
„	„ lanceolaria	Kajgolitsa-yel.
„	Leea sambucina ....⎫ Syn. Leea staphylea ....⎭	Dhindi, Dindi.
18 Sapindaceæ	Hemigyrosa canescens ......⎫ Syn. Cupania canescens...⎭	Karpá.
	Schleichera trijuga	Kusimb, Kosamʰ, Kosham.
19 Anacardiaceæ......	Mangifera indica	Amba (Mango).
20 Connaraceæ.........	Connarus monocarpus	Sundar.
21 Leguminosæ	Crotalaria Leschenaultii	Dνli Dingala.
„	„ retusa	Ghágri.
„	Erythrina indica	Pángára, Páramga.
„	Butea frondosa°	Palas, Khákra, the " Flame of the Forest" (Sanskrit, Palása).
„	Flemingia strobilifera	Boudar.
„	Dalbergia latifolia.............	Sisu, Sawa, Sisam, Táli. (Blackwood Tree).
„	„ volubilis	Alεi.
„	„ paniculata............	Phánsa.
„	„ sympathetica	Penduli-yel, Yek-yel.
„	Wagatea spicata	Vagáti.
„	Mezoneuron cucullatum	Rági.
„	Cassia fistula	Báháwa, Garmala. (Indian Laburnum).
„	Bauhinia racemosa	Apta.
„	„ malabarica..............	Kanchan.
„	Acacia concinna................	Chikakai, Shikikai.
„	„ Catechu†	Kher.

* The leaves of the Palas tree are given as fodder to buffaloes. The flowers are made, with alum, into the yellow dye used at the *Holi* festival (Brandis). This tree gives its name to the memorable plain of *Palasi*, vulgarly called " *Plassey*" (Birdwood's *Vegetable Products*). It yields a *kino* and a *lac.—(Ib.)*

† *Catechu* is manufactured in the Konkan from the wood of the Kher tree·

Natural Order.	Genus and Species.	Vernacular Name.
21 Leguminosæ	Albizzia stipulata	Lullei, Laeli.
,, ...	,, amara	Siras (near Alexander Point).
.. ..	Vigna vexillata	Pirambol, Halula (Matheran Sweet Pea).
,, ...	Cylista scariosa	
22 Crassulaceæ.........	Bryophyllum calycinum... ⎱ Syn. Kalanchoe pinnata... ⎰	Pánphue
,,		
23 Rhizophoreæ	Carallia integerrima	Phansi.
24 Combretaceæ	Terminalia belerica	Behẹrá. Yẹlá.
,, ...	,, Chebula	Hirda (Chebulic Myrobolan Tree).
,, ...	,, Arjuna..	Ain.
,, ...	Combretum ovalifolium	Mái-vel.
,, ...	Calycopteris floribunda...... ⎱ Syn. Getonia floribunda... ⎰	Bẹgvel, Yakshi.
25 Myrtaceæ............	Eugenia Jambolana .... ⎱ Syn.Zizygium Jambulanum ⎰	Jámbul, Jámbu (the common Jambul tree).
	Careya arborea	Kumbhá.
26 Melastomaceæ......	Memecylon edule	Anjan (Iron-wood tree).
27 Lythraceæ	Lagerstrœmia parviflora	Naneh.
,, ...	,, fios regina	Taman.
,, ...	Woodfordia floribunda...... ⎱ Syn. Grislea tomentosa... ⎰	Dhauri.
28. Onagraceæ	Ludwigia parviflora	
29 Samydaceæ.........	Casearia graveolens	Bokhárá.
,, ...	,, esculenta	Mori.
30 Cucurbitaceæ	Trichosanthies palmata.........	Kaúndel.
,, ...	Cucumis trigonus	Kat-vel.
31 Begoniaceæ	Begonia crenata................	
32 Umbelliferæ	Hydrocotyle asiatica °	Brahmi,Karivana,Khopri
,, ...	Peucedanum grande .... ⎱ Syn. Pastinaca grandis ... ⎰	Bápbli.
33 Loranthaceæ	Loranthus involucratus	Bandguli.
,, ...	,, lonicereides	⎱
,, ...	,, lageniferus	Bánda, Vánda
,, ...	,, cuneatus -	(Parasitic plants).
,, ...	,, Wallichianus	⎰
,, ...	Viscum angulatum..............	(Indian Misletoe.)
34 Rubiaceæ	Mussœnda frondosa	Bhút kes, Sárwad.
,, ...	Randia dumetorum	Gela.
,, ...	Canthium umbellatum	Arsul, Tupa.
,, ...	,, angustifolium	Ubáp-yel.
,, ...	Vangueria edulis...	Alu (Indian Medlar).
,, ...	Ixora nigricans..................	Lckbandi, Atkura.
,, ...	Pavetta indica	Pbáphat, Pápat, Pháptí (Matherau Coffee).
,, ...	Adina cordifolia.............. ⎱ Syn. Nauclea cordifolia... ⎰	Hed.
,, ...	Stephegyne parviflora ...... ⎱ Syn. Nauclea parviflora... ⎰	Kalam.
35 Compositæ	Vernonia conyzoides	Moti-sadori, Sahadevi.
,, ...	Cyathocline lyrata..............	Gangotri.
,, ...	Blumea holosericea	Bhamburda.
36 Campanulaceæ ...	Lobelia nicotianæfolia	Dháwal, Devnal.

* An infusion of the leaves of this plant was used by the late Dr. Bhau Daji in his treatment of leprosy.

Natural Order.	Genus and Species.	Vernacular Name.
37 Myrsinaceæ	Embelia ferruginea	Ambati.
38 Sapotaceæ	Bassia latifolia	Mohrá, Máwa, Mohá (Mowrah tree).
,.	Mimusops Elengi	Bokal, Bakúli (below Simpson Lake).
,)	Sideroxylon tomentosum ...	Kánta-kúmbal.
	Syn. Sapota tomentosa... }	
39 Ebenaceæ............	Dispyros assimilis .... }	Malia (Indian Ebony).
	Syn. Diospyros nigricans	
	,, Goindu...............	Goindu.
40 Oleaceæ	Jasminum arborescens (var. latifolium) .... }	Kúsar.
	Olea dioica	Pár-jámbul, Párjám (Wild Olive).
41 Apocynaceæ.........	Carissa carandas	Karwand, Corinda, (Corinda bush).
,, ...	Holarrhena antidysenterica ...	Kura, Indrajav.
,, ...	Tabernœmontana crispa	Pándhra kura.
,, ...	Wrightia tinctoria	Kála-kura.
,, ...	Anodendron paniculatum	Lambtáni(Dr.MacDonald's "Seed Traveller").
42 Asclepiadeæ	Calotropis gigantea	Rui, Ak, Madár.
,, ...	Gymnema silvestris	Káwali.
,, ...	Hoya pallida	Dudh yel (Wax-plant).
,, ...	Leptadenia reticulata	Khar-khodi.
43 Loganiaceæ	Strychnos colubrina	Kánal, Kájer-vel (near Simpson Lake).
,, ...	,, potatorum............	Niwali, Nirmali (near Hart Point).
44 Gentianaceæ	Exacum pumilum...............	Ja'áli.
45 Boragineæ	Paracaryumcœlestinum......	Nechurdi.
	Syn. Cynoglossum cœlestinum... }	
46 Convolvulaceæ ...	Argyreia sericea............	Gavel.
47 Solanaceæ	Solanum indicum............	Obiturti, Bhui-vángi.
48 Bignoniaceæ	Heterophragma Roxburghii ...	Wáras.
,, ...	,, adenophyllum	Pádel.
49 Acanthaceæ	Thunbergia fragrans	Eri-yel.
,, ...	Strobilanthes asperrimus	Kárvi (Indian Wattle).
,, ...	,, Heyneanus	Akrá.
,, ...	Blepharis asperrima	Pahádi-stgan.
,, ...	Haplanthus verticillaris	Kálá-kirát, Kála-ákra.
,, ...	Barleria strigosa............ }	Koranta.
	Var. terminalis.	
,, ...	Barleria coartallica	Itári.
,, ...	Hygrophila serpyllum }	Rán-tewan.
	Syn Physiohilus serpyllum	
,, ...	Ecbolium Linneanum }	Dhákta-adulsa.
	Syn. Justicia Ecbolium...	
,, ...	Phaylopsis parviflora }	Wáiti.
	Syn. Aetheilema reniforme	
50 Verbenaceæ	Callicarpa lanata }	Yesur, Eshwar.
	Syn. Callicarpa cana	
,, ...	Tectona grandis............	Ság, Ságwan (Teak tree).
,, ...	Premna coriacea............ }	Chámbár-vel.
	Syn. Premna scandens.....	
,, ...	Gmelina arborea............	Shewan.
,, ...	Vitex Negundo	Negud, Nirgundi.

Natural Order.	Genus and Species.	Vernacular Name.
51 Labiatæ	Pogostemon parviflorus Syn. Pogostemon purphricaulis	Páugla, Pángli:*
„	Colebrookia ternifolia	Bháman.
„	Anisomeles Heyniana	Chodbárá.
„	Leucas stelligera	Gumá.
52 Chenopodiaceæ	Chenopodium ambrosoides	Danni.
53 Thymelaceæ	Lasiosiphon speciosus	Rametta.
54 Lauraceæ	Machilus glaucescens	Gúlúm.
„	Actinodaphne lanceolata	Pisbá
55 Elæagnaceæ	Elæagnus Kologa	Ambulgi.
56 Piperaceæ	Piper silvestris	Dongri mirchi(Hill Pepper).
57 Euphorbiaceæ	Tragia involucrata	Kulti (Sting-nettle Creeper).
„	Macaranga Roxburghii	Chandárá.
„	Rottlera tinctoria	Robin, Roen, Kapila.
„	Croton hypoleucor	Pandurai.
„	„ Lawianus	Borambi.
„	Briedelia montana	Asána:
„	Phyllanthus madraspatana	Kanocha:
„	Ceratogynum rhamnoides	Chikli.
„	Emblica officinalis	Awala (Gooseberry tree).
„	Glochidion lanceolarium	Bhoma.
„	Fluggea leucopyros	Pándhar-phali.
58 Ulmaceæ	Sponia Wightii	Gol.
59 Urticaceæ	Fleurya interrupta	Khájoti.
„	Gerardina heterophylla	Moti-khájoti, Agia.
60 Moraceæ	Urostigma cordifolium	Pábir.
„	„ retusum	Nandruk, Raneknit.
„	„ religiosum Var. (?)	Ashta † (Sanskrit, Ashwath).
„	„ infectorium	Kel.
„	„ bengalense	Wad (Banyan tree; below Chowk Point.)
	Covellia glomerata	Umbár (the "Sycamore tree" of the Bible).
„	Ficus heterophylla	Ditir.
„	„ oppositifolia	Kharoti.
61 Artocarpaceæ	Artocarpus integrifolia	Phanas (Jack tree).
62 Guetaciæ	Gnetum scandens	Umli.
63 Smilaceæ	Smilax ovalifolia	Got-vel.
64 Dioscorineæ	Dioscorea pentaphylla	Shend vel.
„	Discorea bulbifera	Karu-karanda, Nor-vel.
65 Liliaceæ	Chlorophytum breviscapum	Kula.
66 Aroideæ	Arisæma Murrayii	Sámpatsa-khánda(Snakeroot, the "Cobra Lily").
„	Amorphophallus campanulatus.	Suran.
„	Remusatia vivipara	Rekh-álu.

* The leaves of the Pángli are believed in the Konkan to be a cure for snake-bite. A case of an alleged cure was lately brought to the notice of the Revd. Fr. Dreckmann in Bombay. A man had been bitten by a poisonous snake and was said to have recovered after the application to the wound of the leaves and other parts of two plants, which were produced; and one of these was apparently the Pángli.

† The Ashta is distinguished by the Hill-people from the Pipal of the plains, of which it is perhaps a variety.

Natural Order.	Genus and Species.	Vernacular Name.
67 Orchidaceæ	Aerides crispum	
,,	,, maculosum	Ichwáoh,
,,	Dendrobium barbatulum	
,,	,, chlorops	
,,	Platanthera Susannæ?	Kálábi,
,,	Habenarja candida	
68 Burmanniaceæ	Burmannia triflora	(On the road to the Governor's bund.)
69 Musaceæ	Musa ornata	Rán-kel, Kawadar, Chá-wan-kel(WildPlantain)
70 Zinziberaceæ	Curcuma Zedoaria	Kachora, Kachola.
	,, pseudo montana	Rán-haldi (Wild- Turmeric).
71 Amaryllidaceæ	Pancratium parvum	Khandálú.
72 Hypoxidaceæ	Curculigo malabarica	Kajuri.
,,	,, graminifolia	
73 Palmæ (Palms)	Caryota urens	Bherli-már° (Fish-tail Palm).
74 Gramineæ (Grasses)	Coix lachryma	Kasai, Rán-makai (Job's Tears).
,,	Bambusa stricta	Váns, Bámbu (Bamboo).
,,	Andropogon muricatus,	Wals-khaskhas (Khaskhas Grass).
	Andropogon?	A Grass, with the smell of turpentine, near the Neral Station.
75 Filices (Ferns)	Sagenia coadunata Syn. Aspidium cicutarium.	Kájáryatee Báshing (Indian Beech Fern),
,,	Asplenium planicaule Syn. Asplenium laciniatum.	
,,	Pteris aquilina	Brake Fern.
,,	,, quadriaurita.	
,,	,, pellucida	
,,	Adiantum lunulatum	Hansráj, Rájhana (Goosefoot Maidenhair Fern).
,,	Cheilanthes farinosa	Pátkuri (Silver Fern).
,,	Lygodium pinnatifidum Syn. Lygodium flexuosum.	Hansráj yel (Creeping Fern).
,,	Polybotya appendicula'a	
,,	Acrostichum virens Syn. Pæcilopteris virens, Gymnopteris contaminans.	Rooting Fern.
	Polypodium quercifolium Syn. Drynaria quercifolis.	Kadik-pan (Indian Oak Fern).
76 Lycopodiaceæ (Club-mosses)	Lycopodium imbricatum	
77 Musci. (Mosses)	Hypnum curratum	
,,	,, squarrosum	
,,	,, bryoides	
,,	,, reflexum	
78 Fungi	Agaricus campestris	Alamben (Mushroom).
,,	Lycoperdon pratense	Bhoipbor (Puff Ball).
,,	Dœdalia gibbosa	Kerambi, Páranza.
,,	,, versicolor	
,,	Polyporus giganteus	

* The long trailing fruit racemes of this Palm are likened by the Hill people to the flowing locks of the long-haired Bheravs, attendants of Shiv: hence the name "Bherli-mar."

INDEX OF VERNACULAR NAMES.

(The references are to the figures in the first column of the Catalogue.)

A LIST OF THE BUTTERFLIES OF THE BOMBAY PRESIDENCY IN THE SOCIETY'S COLLECTION.

With Notes by E. H. Aitken.

THE two following species were omitted by me in the first part of this paper which appeared in July. I have nothing to note about either of them:—

39. *Mycalesis mineus.*—There is a single specimen in the collection, without locality.

40. *Ypthima singâla.*—This also is without note of locality. The Society's collection is rather weak in Satyrinæ.

I find that I also omitted to mention that in Bombay I have found the larva of *Junonia limonias* on *Barleria prionitis,* a near ally of *Asteracantha longifolia,* the favourite food of *J. almana.*

I will now proceed with my list.

LEMONIDÆ.

41. *Abisara fraterna.*—When the rainy season is drawing to a close, in September or October, every bush on the hills is enlivened by the attitudes and frolics of this little embodiment of vanity. In all its ways it is unique, perching in the middle of a leaf, on the upper side, with wings half open, turning jerkily from one side to another, then hopping to another leaf and strutting round it. Sometimes a pair join in these performances, It is one of the easiest Butterflies to catch, having no fear. I have found it in Poona, but rarely, if ever, in Bombay, though it is common in the low jungles of the Tanna District.

LYCÆNIDÆ.

I divide the Lycænidæ by form and habits into two strongly contrasted tribes; the one, robust in body and brilliant in colour, swift and wary, given to basking on high trees, may be illustrated by such genera as *Virachola* and *Tajuria*; the other, a feeble folk, without character, flitting mostly near the ground, or resting on low bushes with their wings very slightly opened, includes such genera as *Catochrysops, Polyommatus, Zizera,* and their kindred. The former have the thorax very stout, few Butterflies comparing with them in this respect, except the species of *Charaxes* and some *Hesperidæ*; but they pass gradually into the weaker forms through such genera as *Aphnæus,* and, as I do not propose to be the founder of a new classification, I will merely place the genera in such order as seems best to illustrate my idea.

42. *Anops phœdrus.*—This little gem, though nowhere plentiful, may be met with in every part of the Presidency. It appears after, or perhaps before, the end of the monsoon, and remains till the end of the year. In the afternoon, when most other Butterflies have retired to rest, it loves to bask in the sun on a small tree or high bush, with wings just a little open.

43. *Baspa melampus.*—I have not often caught this, which is rarer than the last, and have seen it too seldom to form an opinion on the regular time of its appearance. It has the same habits as the last.

44. *Tajuria longinus.*—This also is comparatively scarce, but occurs, I think, almost everywhere.

45. *Irasta marunas.*—I do not think I ever caught with my own hands this most splendid, surely, of all the Lycaenidæ, and I doubt if it occurs in Bombay. The specimens in the Society's collection are all, I think, from the Tanna or Nasik District, and I have met with it myself at Egutpura on the Thull Ghât, where it began to appear in October or November.

46. *Virachola isocrates.*—It is almost impossible with the net to get a really good specimen of this or the next. They are not only difficult to catch, being exceedingly swift, wary, and given to settling on high trees, but, when caught, difficult to secure without injury. There is a delicate bloom on a fresh specimen which the gentlest touch destroys. It is easily reared however. As is well known, the larva feeds inside the fruit of the pomegranate and, some time before becoming a pupa, eats its way through the tough rind and fastens the fruit with silk to its stalk, thus preventing it falling off in case it should wither before the Butterfly escapes, as it generally does. This operation is performed at night, and generally repeated night after night. I have taken a pomegranate infested with these larvæ (several usually inhabit each fruit) and made it stand in an egg-cup; in the morning it was so securely fastened that in taking up the fruit I lifted the cup. Of all animal instincts that I have seen or heard of, this is one of the most astonishing and certainly the most difficult to reconcile with any theory of development. As far as I have observed it, the larva never leaves its shelter except for the definite purpose so necessary to its safety, and it taxes ordinary ingenuity to suggest any possible conditions under which some larvæ might have performed the act in the first instance without purpose. I have found this Butterfly pretty common in Bombay and Poona from December or January till March at least.

47. *V. perse.*—I do not think I have met with this except on the hills, where it is common, appearing in December when the fruit of the Ghela (*Randia dumetorum*), on which the larva feeds, is ripening, and remaining till March or April. The larva has the same curious instinct as the last species and needs it more, for the Ghela fruit withers at once when attacked and would inevitably fall before its tenant had reached the pupa state if not artificially supported. I have found only one larva in each fruit, and have sometimes noticed ants going in and out of the hole made by it, for what purpose I cannot say. The stony hardness of the fruit turns the edge of one's penknife and of one's curiosity too. This Butterfly has the habit of taking its station, during the hottest hours of the day, on

a particular leaf, from which it darts out in pursuit of every other Butterfly that passes by. This habit characterises a few brilliant genera in families widely different. It is strong in Charaxes.

48. *Nilasera amantis.*—This is not common, and I am not sure of the limits of its season. I have seen it oftener about the beginning of June than at any other time, and oftener at Karanja across the Bombay Harbour than at any other place. It flies very fast.

49. *Aphnæus* (or *Spindasis*) *vulcanus.*—This species is not to be met with in Bombay gardens; but in the Deccan it is not rare, and on Karanja I have found it abundant in the hot season. I think it rarely opens its wings, except to fly.

50. *A. acamas.*—Mr. Newnham sent specimens of this from Bhooj.

51. *A. trifurcata.*—These are without note of locality, and I know nothing of them.

52. *A. elima.*—These are without note of locality and I know nothing of them.

53. *Catapæcilma elegans.*—A single specimen of this was caught by Mr. R. C. Wroughton at Bassein in the Tanna District last March or April.

54. *Rahinda amor.*—This occurs almost everywhere, but is common, nowhere. It appears at the close of the rainy season. It is fond of taking its stand on the point of a prominent leaf, with wings closed and an air of decision not easy to describe. *Spindasis* has the same habit.

55. *Jamides bochus.*—This is not uncommon in Bombay and the surrounding country, and also in Poona, chiefly, I think, after the monsoon, but I have no notes.

56. *Tarucus theophrastus.*—Common both in Bombay and the Deccan after the rains. Specimens vary much in size and in the intensity of the spots on the under side. The larva feeds on the tender leaves of the Beyr or Bor tree (*Zizyphus jujuba*).

57. *T. plinius.*—This is not so common as the last, but not rare coming out at the same season. I have found the larva on *Sesbania aculeata*, an annual which springs up everywhere in Bombay during the rains and shoots up to a height of 6 or 7 feet and withers away in October. Its fragile leaves wither up a few minutes after being plucked, and it is no easy matter to rear a minute larva on them. I was successful with only one. I find it described in my notes as green and of the usual wood-louse form, with a dorsal ridge of small protuberances. The pupa, which came out in seven days, was greenish, smooth, not ¼th of an inch long, and closely attached to the bottom of the pill-box in which it was kept.

58. *Castalius rosimon.*—Very common from August to the end of the year at least, alike on the hills and the plains. It settles much on the ground.

59. *C. decidea.*—I believe, but am not quite certain, that I have caught this in Bombay. It is not uncommon on the hills.

60. *Talicada nyseus.*—This peculiarly distributed insect is not found at all in Bombay, nor do I recollect once meeting with it at Khandalla, Matheran, or Egutpura; but in a particular spot at Mahableshwar it was swarming last March, and I have a faint recollection of its being equally abundant at the hill forts of Singhur and Poorundhur near Poona, while at Poona itself it is never wanting during the dry months. Mr. H. Wise informs me that in Kanara he finds it at an elevation of 1,500 feet. It lies very low and settles much on the ground, wings always closed.

61. *Lycaenesthes lycaenina.*—There is one specimen, a male, in the collection, without note of locality. I have a strong impression that I myself caught it in Bombay and forgot to label it at the time.

62. *Lampides œlianus.*—This is not confined to the hills, but decidedly more abundant there than on the plains. About Christmas there is no insect more abundant at Khandalla.

63. *Catochrysops cnejus.*—This is very common everywhere after the monsoon. There is little to note about these commoner Lycaenidæ. They are very much alike in their ways, flying low and often basking with their hind wings more expanded than their fore wings, a habit which they share with some of the Hesperidæ. Some of them have also the curious habit of rubbing their hind wings against each other.

64. *C. strabo.*—This appears also after the monsoon, about August, but is not so common in Bombay, I think, as the last.

65. *Polyommatus boeticus.*—This is common everywhere.

66. *Chilades varunana.*—There are five specimens in the collection without note of locality, but certainly from the Tanna or Nasik District I know nothing about it.

67. *Pathalia albidisca.*—There are a few specimens from different parts of the Presidency.

68. *Azanus crameri.*—A single specimen without note of locality.

69. *Spalgius epius.*—I have found this on Karanja in February, August, and September, but it is not common.

70 *Zizera karsandra.*—I find myself obliged, with shame, to confess that I am not quite sure whether this is the species which swarms all over the Esplanade in Bombay some time after the rains. I assumed that I knew it, and now, when a doubt has arisen in my mind, I am no longer in Bombay. It can scarcely however be any other species.

71. *Z. pygmaea.*—This is a Bombay species too, but not so abundant.

72. *Z. ossa.*—This has been described by Colonel Swinhoe for the first time in the paper which I have already referred to. It is not by any means uncommon.

E. H. A.

ZOOLOGICAL NOTES.

NOTE ON THE *HOMALOPSIDÆ* IN THE SOCIETY'S COLLECTION.

By Mr. James A. Murray, *Curator, Karachi Museum.*

In August last I had the pleasure of examining a good part of the Society's collection of reptiles, and among them the specimens (six in number) of the Homalopsidæ, described in No. I of the Society's Journal by the Rev. F. Dreckmann. The specimens were correctly referred to the Homalopsidæ, but were not assigned to any group evidently owing to the difference in the number of scales found the body. The other characters agreed quite with those of the genus *Ferania*, and I had no hesitation in identifying the specimens as *Ferania Sieboldi*, (Schbg.,) on finding that the specific characters of the only species known also agreed. When Dr. Gray founded the genus *Ferania* (*Zool. Misc.*, p. 67), he had but a single specimen from Province Wellesley in Bengal, and one with only twenty-seven series of scales round the body. Lieutenant Barnes has done good service in unearthing several more specimens, and thus being the means of bringing about an amendment of the generic characters of one of the four genera, constituting the group of Homalopsidæ, having no nasal appendage and more than five upper labials. The generic characters of *Ferania*, as now amended, will shortly stand as under :—

Snout without appendage ; more than five upper labials ; *two* anterior frontals ; scales in 27—31 series.

One species, *F. Sieboldi*, (Schbg.,) characters as described in Gunther's *Reptiles of British India*, p. 284 ; scales in 27—31 series.

J. A. M.

LIST OF BUTTERFLIES RECEIVED FROM MAJOR YERBURY,

Campbellpur, Punjab.

5	Hipparchia parisatis	2	C. sareptensis.
2	Aulocera swaha.	3	Euchloe lucilla.
1	A. saraswati.	2	Mancipium canidium.
2	Amecera schakra.	3	M. nipalensis.
1	Callerebia daksha.	2	Catopsilia pyranthe.
2	Ypthima asterope.	2	Teracolus faustus.
2	Y. bolanica.	1	T. fimbricata.
2	Y. nareda.	5	T. protracta.
1	Danais limniace.	1	T. etrida.
1	Vanessa cashmiriensis.	2	Deudorix epijarbas.
2	Junonia asterie.	1	Baspa nissa.
1	J. orithyia.	6	Spindasis acamas.
1	Argynnis niphe.	3	Catochrysops cnejus.
1	A. lathonia.	1	Tarucus nara.
4	M. Robertsii.	1	Lycæna putli.
2	Libythea lepita.	3	Zizera maha.
2	Dodona durga.	2	Z. trochilus.
1	Papilio erithonius.	1	Chrysophanus phlæas.
1	Belenois mesentina.	1	Hesperia evanidus.
2	Gonypterix nipalensis.	5	Gegenes karsana.
5	Colias Fieldii.		

NOTE ON THE CONDUCT OF A TAME PIGEON.
By E. H. Aitken.

THE curious example of conjugal infidelity among pigeons given by Mr. Hart in the last number of the Journal reminded me of two incidents, illustrating the characters of the same birds as husbands and fathers, which may interest members. By way of parenthetical preface, I will say that, if the Journal of the Bombay Natural History Society awakens a livelier interest in the behaviour of animals as intelligent beings, it will do a valuable work.

In 1879 a baby pigeon, not more than a week old, in one of the nests in my pigeon-house, was left an orphan by the sudden death of its mother. It was too young to be fed by hand and I supposed it must die, but I was mistaken. The bereaved father, instead of giving himself up to sorrow, at once took sole charge of his helpless offspring and reared it successfully. He had not sense to make any change in his habits. Among pigeons the female sits alone on the nest, except for three or four hours in the middle of the day, when she is relieved by the male; so this bird went in every day, about 10 or 11 o'clock, and kept the nest warm till 2; but all night he slept as he had been accustomed to do, in another chamber, leaving his naked little child exposed to the cold of a February night. It survived however and was doubtless all the hardier for its Spartan nurture.

Whether this parent's conduct is attributed to intelligence or stupidity will depend upon the direction in which we have accustomed our feelings to run; but there can be no question about the following case. In my flock there was one old male bird who was quite a character in the community. He was a fat easy-going, good natured bird, but pampered and self-indulgent to an uncommon degree. It was a favourite sport of mine to fit him into the mouth of a stone jar, like a cork, only his head and shoulders out, and in that position to give him grain, which he would eat with the most composed enjoyment. His wife was a blue rock with all the strong instincts and affections of a wild bird. Finding her always willing to take more than her share of the family cares he shirked his and, during the hot season, gave up taking his turn on the nest altogether leaving her to sit day and night, which she did, excepting a very short interval which she allowed herself for food. When the cold season came round, he found his opportunity to repay her by taking all the night work duty on himself. He actually turned her off the eggs and slept in the nest himself, while she roosted at the entrance and kept out the cold air!

E. H. A.

NOTE ON <i>DANAIS DORIPPUS.</i>
By Mr. A. T. H. Newnham, S. C., 10th N. I.

Mr. Aitken mentions in his paper on Bombay Butterflies that he has never met with this variety, but in the last month I have seen here, in Cutch, two specimens, one of which I added to my collection. Besides these, another collector obtained two more at Mandvie, and said he had seen others which escaped him. Also the same collector had caught the variety known as <i>D. alcippoides,</i> but having the lower half of the hind wings pale lavender seaintd of white.

A. T. H. N.

NOTE ON LOCALITY.

By Mr. A. T. H. Newnham, S. C., 10th N. I.

Extraordinary Coincidence.—During a recent visit to Ceylon I happened to go again to a certain spit of shingle on which I had a month previously found several eggs of *Sterna melanogaster*. I was again successful in finding two eggs of the above-mentioned bird, and on lifting the eggs up to deposit them in cotton-wool, my eye was caught by something glittering on the spot from which I had just removed the eggs. On picking it up, I found it to be an " entomological pin," and presumably one which I dropped when I was there before, as it is in the highest degree improbable that any one else would have had entomological pins in such an out-of-the-way place. The question arises, was it a mere coincidence that the Tern laid its eggs on that very spot, or was it attracted by the glittering appearance of the pin?

The Bower-bird of Australia, I believe, collects gaily-coloured and glittering objects and places them round about its nest. Could then this Tern have been actuated by some similar freak, and have brought the pin from some place where it had found it?

A. T. H. N.

NOTE ON THE BREEDING OF *PARRA INDICA*.

By Lieut. H. Edwin Barnes.

Mr. Hume in his *Nests and Eggs of Indian Birds* lays stress upon the alleged fact that the Bronze-winged Jacana lays a much greater number of eggs than its nearest Indian ally, the Pheasant-tailed Jacana (*Hydrophasianus chirurgus*).

At page 591 of the above-quoted work, Mr. Hume writes :—" Of six nests examined, none contained more than *seven*, but the boatmen averred that the birds, sometimes at any rate, laid *ten*."

Again, on the next page, quoting from Mr. R. Blewitt's experiences in the Jabulpur, Saugor, and Jhansi Districts, he writes :—" The regular number of eggs I have not been able to ascertain accurately, but from *eight* to *ten* may be taken as the maximum number."

I have had opportunities of examining great numbers of these nests *in situ*, and I have never yet found more than *four* eggs in any one of them, although many have been in an advanced stage of incubation ; the fishermen, too, assert that four is the number invariably laid. I cannot help suspecting that some mistake has occurred. I actually took with my own hands over two hundred eggs, on four different dates, in August and September 1880, from jheels in the vicinity of Neemuch, and I have taken at least fifty eggs from the Saugor and Chundrapur Lakes this season, and had I wished could easily have taken four times as many. The Saugor Lake is within half a mile of my bungalow, and is much frequented by these birds, and as I am continually boating and fishing upon it, I have exceptional opportunities of noting facts in reference to their habits and nidification.

I cannot help coming to the conclusion that four is the normal number of eggs laid by this bird, and that whenever a greater number has been found, it is the joint production of two or more birds.

I do not remember seeing the fact noticed anywhere that these birds often deposit their eggs on a heap of floating weeds without preparing any nest at all.

It would be interesting if other zoologists would state if their experiences coincide with mine or not.

H. E. B.

NOTE ON REVERSION TO PRIMITIVE TYPES.

By R. A. STERNDALE.

I HAVE mentioned in the *Mammalia of India*, quoting from a writer in the *India Sporting Review*, a case of cross-breeding between jackals and dogs, in which in the third generation, or one-eighth jackal and seven-eighths dog, three out of five pups had gone back to the jackal type. I have since then been noticing cases of reversion in domestic cats. We have an English, or rather Scotch, black cat which we brought out from home three years ago. Her first kittens in India were all white, with patches of the usual Indian grey or Indian tabby, which consists of small spots in lines on a grey ground. We destroyed all except two, a son and daughter, the latter a very pretty cat, with decidedly English points about her : this cat, in her third family of the usual grey-and-white kind, had one very handsome tabby kitten, which, with a white one, was kept. Now this tabby kitten, who was named, "Joe," because like Dickens' fat boy, he was always sleepy, afterwards softened into "Joey," turned out a true English tabby, a type I have never seen in India (*see* the sketch I have given of him in this journal), and a tabby of a very handsome kind, unusually so. Were he to escape in suitable jungles and be shot, he would probably, but for his tail, be identified as *Felis marmorata*, for he is nearer in colouring to that species than any domestic cat I have come across. Even pure English tabbies have, like their remote ancestor the wild cat (*Felis catus*), certain stripes down the sides, but Joey, with the exception of the bars on his limbs, is clouded like the Rimaudabau (*Felis diardi*), or the smaller marbled cat (*F. marmorata*). English tabbies do occasionally have their side markings in irregular concentric circles, but the colour of the ground-work is generally grey instead of sandy fulvous. However I take it that Joey gets his Joseph's coat of many colours from his English ancestor and not from his Asiatic grandfather. He is a queer-tempered cat, shy with most people, although his sisters and his cousins and his aunts will go to anybody ; but he is devoted to me, and at times will not leave me for a moment. Lately, whilst laid up with the fever which has delayed the issue of this journal, I had to keep to bed for a day or two and Joey never left my side, and his meals had to be brought into my room.

Now to go back to Joey's grandmother, the old black Scotch cat. For two years-and-a-half she had a constant succession of grey-and-white kittens' between twenty and thirty, and we wondered why none of her children resembled her. Lately however, out of a batch of five, three were jet black.

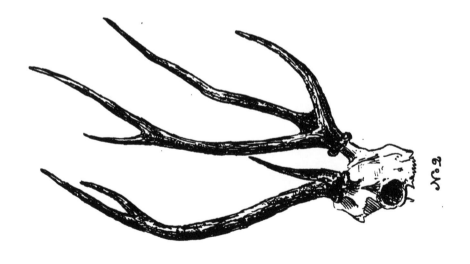

No. 2

Her eldest daughter (Purry) had lately five kittens, of which one was jet black and two others partly so; but it is only recently that this colour has begun to shew itself: of some thirty or forty preceding kittens, only two had a few black patches—now blackies are getting common.

I forgot to state that Joey's markings are perfectly symmetrical, each side being alike. As you hold him up with his back towards you, the pattern runs off on each side from the central stripes as evenly as if they had been marked off with compasses, which is characteristic of the feræ.

R. A. S.

SOME NOTES ON ABNORMALITIES IN THE HORNS OF RUMINANTS.

By Mr. J. D. Inverarity.

THE most curious instance of abnormal horns I have met with was an old stag samber I shot some years ago near the Taptee. His right horn was 36 inches long and nothing peculiar about it. The left horn was a few inches shorter, and had no brow antler at all nor the slightest rudiment of one. About three-quarters of an inch from the left horn was a third horn, a mere knot but growing *on a separate bony pedicle of its own*. It was entirely distinct from the main horn, the skin covering the intervening space. No sign of disease or injury to any of the organs.

SINGLE SAMBER HORN (Sketch No. 1).—A very massive heavy horn. Either shed or killed by tiger. I think the latter. The horn had the appearance of having dripped over all round the burr and hung down in what, for a better term, I will describe as numerous icicles. This horn, for months after I picked it up, sweated some oily matter of a most offensive odour.

FOUR-HORNED ANTELOPE.—The bony core curves inwards of one horn. Had anterior horns, but I have lost them.

WILD COW-BUFFALO.—Right horn normal, about 3 feet long; left horn not more than 18 inches long, probably less, growing almost straight down close to the cheek and turning backwards. I was close to her for several minutes and observed it well, but did not fire at her. The misshaped horn appeared much thinner and smoother than the other one. The end *was blunt*.

There is a curious malformed cow-bison's head in the Madras Museum, of which you might get a sketch.

I have in my possession in Scotland a small samber head the left horn of which bends down, forming a club close to the skull just like the horn of the Cashmere stag pictured No. 2. This club shape is the natural shape of deer's horns while growing. Any one who has seen stags while their horns are growing, before they have reached the point where the upper tines branch out, will corroborate me that the top of the horn is then club-shaped. The Cashmere stag No. 2 and the small head I am speaking of have had their growth arrested at this stage.

DOE CHINKARA.—I have never shot one, but I think their horns are frequently misshapen. One I have a note of had one horn bent forward and the other backwards, but having omitted to take a drawing of it my note is not sufficiently full to enable me to give a more accurate description.

See too the curious bison head of mine, shot in 1885, in the Society's Rooms. The bony core being only a few inches long, there was nothing to give the usual bend to the horns, which have accordingly grown straight out and curved forward. This was a very old cow, the incisor teeth being worn level with the gums—a thing I have never seen before. I shot her by a fortunate accident. There were a lot of three bison. I noticed something peculiar about the head during the stalk, but did not see what the real state of the case was. Firing at the large bull, I broke his shoulder. The second barrel was intended for the bull, but the cow rushed alongside as I pressed the trigger and got the bullet in the neck, dropping dead.

I also send for inspection a small samber head. I am not sure whether the right horn has ever had a brow antler. There has been a fracture of some sort. Whether the brow antler has been broken off and the fracture worn smooth, or, as I am induced to think from there being no fracture visible on the inside of the horn, that there was no brow antler, is doubtful. If the latter is the case, the long brow antler (for the size of head) of the left horn is remarkable.

PTEROPUS EDWARDSI.

I saw on 9th May this year at Nara, on the banks of the Jouk River, a number of Flying-foxes fanning themselves in the way described by Mr. Aitken. The fanners however were only about 10 per cent. of the population.

J. D. I.

Editor's note on above.—Mr. Inverarity was kind enough to send me the above notes to help me in a continuation of my previous paper on horns; but ill-health has prevented my taking up the subject more fully this time so I have published his notes without any addition of my own. I have copied his sketch of the very curious samber horn he picked up ; and have also to thank him for the loan of a book on sport in Madras by "the Old Shikarry" (G. A. R. D.), in which is a photograph of a cheetal's head with an abnormal bez-tine of extraordinary length. I have taken the liberty of copying this, and it forms No. 2 sketch in the accompanying plate.

R. A. S.

NEOMERIS KURRACHIENSIS — (Murray).—The following description of the Porpoise, mentioned in the paper on the *Waters of Western India,* page 159, of which I have given an illustration, has been sent to me by Mr. Murray, and is in fact a draft of his paper on the subject in the *Ann. and Mag. Nat. History,* Vol. XIII., 1884. It will interest our readers and supplement KESWAL'S description.

R. A. S.

"A cætacean of the family Delphinidæ, which I shall describe under the name *Neomeris kurrachiensis.* The characters of the génus are :—Dorsal fin none ; nose of skull short, rounded in front, flat and shelving above ; teeth numerous, compressed, nicked, acute, extending nearly the whole length of the jaw (Gray, 'Seals and Whales', &c.).

"*Neomeris phocaenoides* is the only species of the genus, and its dentition is given as $\frac{14}{15}$ (*Delphinus melas*) or $\frac{20}{19}$ on each side. The species under notice has $\frac{18}{11}$ on each side, and there are besides a set of $\frac{2}{2}$ which were scarcely visible through the gums, and situated out of the line of the other teeth in front of the jaws. In shape these $\frac{2}{2}$ teeth are quite unlike the rest, being conical instead of flattened or compressed. The measurements of the animal taken in the flesh are as under :—

	Inches.
Length along curve from tip of snout to notch between caudal flukes	52
Ditto straight	45
Tip of snout to pectoral fin	10
Caudal flukes	9 × 3
Distance of blow-hole from tip of snout along curve	6·5
Ditto from angle of mouth to eye	1·62
Vent from root of caudal fin	14

Snout rounded; head very convex, rising posteriorly high to the dorsal surface; blow-hole semi-lunar; back with a longitudinal band of spinous tubercles on the vertebral area, beginning nearly opposite the root of the pectoral, widening to 1·5 inch about the middle, and again contracting and ending narrowly opposite or in line with the vent; no dorsal fin; pectoral subfalcate; teeth $\frac{18}{15}$; colour shining black throughout, except a purplish red path in front of the snout (on the upper lip) and on the throat; intestine 31 feet in length; contents of stomach Crustacea (species of *Penaeus*).

	Inches.
Length of skull over curves to upper edge of foramen magnum	10.
Ditto straight from below	8
Height of skull (vertex of super-occipital)	4·25
Tip of snout to blow-hole	4·25
Ditto to interparietal	6·25
Interparietal to upper edge of foramen magnum	3·75
Across maxillaries	4·75
Across blow-hole	1·5
Length of molar	2 0
Ditto of brain cavity	4·0
Greatest space between occipital condyles (upper)	1·5
Across paroccipitals	3·37
Smallest space between occipital condyles at lower third	1·0
Vertical diameter of foramen magnum	1·75
Breadth across last teeth on each side (upper jaw)	2 5
Ditto ditto (lower jaw)	2·5
Teeth-line in upper and lower jaw	2·5
Length of lower jaw to coronoid process	5·62
Greatest vertical depth of ramus	2·62
Palate	4.0

The super-occipital is sub-globular and very convex above; rostrum short, rounded in front; foramen magnum vertically oval, with the occipital

condyles vertically elongated and convex, wider at their lower third; teeth
small, flattened or compressed, with a sharp sub-crescentic crown, faintly
nicked, and with the middle of their outer and inner sides slightly swelled ;
they are rather obliquely arranged in line, about one-fifth of each succeeding hinder one overlapping its fellow, but not in contact.

BOTANICAL NOTES.

NOTE ON THE *GLORIOSA SUPERBA* (N. O. *LILIACEA*), "SUPERB LILY."

BY MR. FRANK ROSE, P. W. D.

SEVERAL writers have pronounced the *root* of this handsome climbing plant
a *violent poison*, and next to the Wild Aconite (*Aconitum ferox*). I much
doubt the assertion, as I have seen *Brinjaris* using it for medicinal purposes,
and it doubtless has active properties. Native Surgeon Mohideen Sheriff
(Madras) has already removed the doubts expressed by certain of the
Medical Faculty by giving it to his patients, and has himself taken "12-grain
doses *three* times a day." In case an experiment may be wished to be tried,
I send you by this day's post the tuberous root of this shrub obtained from
my garden.

Florists should not lose this opportunity of collecting the roots for next
rains. This ornamental plant flowers early in August, lasting only eleven
days : the petals open with a light green tint, and then gradually assume the
crimson and yellow on the sixth day, when it is then clad in its richest and
gayest colour, after which the whole flower becomes crimson and then fades.

This perenial plant is easily identified. The *root* is bulbous ; *stem*, green
herbaceous ; *leaves*, lanceolate, ending with tendrils or cirrhiferous ; *calyx*, nil ;
and *corolla*, reflex, 6-petalled ; *habitat*, fields and forests. *Willdenow* is said to
have discovered this shrub in 1690. The Indian synomyms are *Nag-dhan*
or *Nat-kabachnag* derived from the Wild Aconite ; *Olot-chandal*, Bengali ; and
Kalaippak-kirhungu, Tamil. In *Balfour's* Botany (Ed. 1854) nothing is said
about this plant.

                                                                F. R.

NOTE ON THE *GLORIOSA SUPERBA.*

BY SURGEON K. R. KIRTIKAR.

WITH reference Mr. Rose's remarks, I may at once state that I am not
personally able to bear testimony to the violently poisonous qualities of the
root of *Gloriosa superba*. I have neither used it medicinally, nor have I seen
any cases of men poisoned by it. Dr. Norman Chevers however, in his work
on Indian Medical Jurisprudence, mentions two fatal cases (pp. 284-285)
in Edition of 1870, and attributes to the root narcotico-irritant

properties. Gribble in his recent work on Indian Medical Jurisprudence is silent on the point.

In his *Forest Flora of British Barmah* (Vol. II., p. 542), Kurz says :—" The Phoongyees often collect the poisonous roots of *Gloriosa* for medicinal purposes." Sir George Birdwood, in his *Bombay Vegetable Products*, says it was first described by Hermann. It is said to be a substitute for Colchicum. In Bapu Gangadhar Joshi's *Nighanta Prakásh*, based on various Sanskrit works onthe use and properties of indigenous drugs, the plant is called " Kalikári," " Kalalávi," or " Khadiyánág." It is said to be destructive of biliousness, pruritus, œdema, intense thirst, colic, &c. It is therefore not unknown as a remedial agent. It is said to be abortive also. It is deserving of a trial as tonic and alterative, especially as Mohideen Sheriff finds it useful in his own practice.

K. R. K.

USES OF THE FLOWER OF *PANDANUS ODORATISSIMUS.*
BY MR. FRANK ROSE, P. W. D.

IN forwarding for identification two samples of the extract from the flower of this tree, known as the "Attar of Keura" and the "Water" (*Kevada-ká-aaraq*) manufactured last year at Aurangabad, Deccan, but which have lost, to a certain extent, their aromatic properties from length of time, and with reference to the very interesting paper on the uses of the tree by Mr. R. A. Sterndale, F.R.G.S. (Journal No. II. for April), I am induced to follow up Dr. Kirtikar's " Notes," and say a few words anent the uses of this achlamydeous flower.

THE FLOWER is certainly of a very fragrant nature, more powerful than any of the Indian Flora, and its perfume is considered to be the richest by the Mahomedan community. The flowers are used for a double purposes *viz.*, scenting wearing apparel and keeping away insects, especially the cockroaches (*Blatta. orientalis*).

PERFUMERY.—The *Aaraq*, or water, is issued extensively by the well-to-do of the Mahomedan class, chiefly in flavouring their drinking-water during the hot-weather by adding a few drops to it. Although it may be palatable to some Europeans in their beverages, confectioneries, *et hoc genus omne*, I am no advocate for it ; but tastes differ, hence my reason for sending a sample for trial.

The *Attar* is prized as much by the Native community as any of Piesse and Lubin's perfumes are by Europeans. A superior kind of *Attar* is exported from Northern India.

SYNONYM.—A respected botanist says that the word *Pandanus* has its derivation from *Pandang* (Malay name of the genus), signifying "Regard," owing to "the beauty of the tree and its exquisite odour." Daniel Olliver, F.R.S., F.L.S., says that the "Screw-pine " derives it, appellation from the Pine-apple order (Bromeliaceæ) owing to the similarity of their foliage. It is also known as the "Caldera Bush" and "Screw-palm," and in Mauritius as the "Variquois Plant." The plant was first recognized in India in 1771 by that great German botanist Willdenow. I find that

Mahadeva (*q. v.*, "Hindu Theatre"—*Malati* and *Mahadeva*) is reported to have sung the praises of the *Kitaki* (Sanscrit) in the following strain :—

> "Faint in the East the gentle moonlight gleams
> "Pale as the Palm's sear leaf, and through the air
> "The slowly rising breezes spread around
> "The grateful fragrance of the *Kitaki*."

In Burmah the plant is known as *Sasava* and in Madras as *Tashan-cheddi*.

HABITAT.—This tree I have seen growing in Southern India and in H. H. the Nizam's dominions : common in the vicinity of Aurangabad, Deccan. Having an excellent fibre, I am surprised that it is not cared for and utilized for rope-making in the Nizam's territory; but, if so, it must be to a very limited extent ; the fibre could be more profitably used also in manufacturing paper. The Japanese cultivate the plant extensively for its odoriferous nature ; similarly Burmah, where the tree grows wild and luxuriantly, could augment her revenue by utilizing it too. The tree is largely resorted to by the Ophidia family.

FODDER.—J. C. Loudon, F.L.S., H.S., &c. (1829), says :—"The branches being of a soft, spongy, juicy nature, cattle will eat them very well when cut into small pieces." I know that the taste is unpleasant, and from the fact of the leaves decaying on the tree—especially in the younger plants, which are within the reach of all cattle—this assertion seems rather doubtful.

Altogether the *P. odoratissimus* is a most interesting and valuable product of the vegetable kingdom ; and it is to be regretted that a tree so very useful for economic purposes—from the root to the flower—is not cultivated and brought into use more largely.

F. R.

FREAK IN A *ZINNIA PAUCIFLORA* OBSERVED AND EXHIBITED

By MR. FRANK ROSE, P.W.D.

[N. O. *COMPOSITÆ* (*ASTERACEÆ*).]

(*Syngenesia*—LINN. Sub-Order TUBULIFLORÆ.)

IT was Mrs. Caroline A White who truly said that "the researches of modern botanists have done much to simplify and popularise a knowledge of the vegetable kingdom ; but there are still sufficient mysteries in the organization, sensation, and self-motive power of plants to afford a wide field for inquiry and experiment ; and the more we direct attention to these charming wonders, the more good we shall be doing to our readers, ourselves, and science."

Apropos of the above, I may as well here state—*en parenthesis*—that certain habits of the animal and vegetable kingdoms are analogous, of which I hope to give some interesting facts in due course. The ancients believed that plants and trees have instinct and vegetable *souls*, and looked upon them as animals! However, be that as it may, there is no doubt that the floral world has its *lusus naturæ* like animals, as will be perceptible in the specimen of the *green* flower of the *Zinnia pauciflora* herewith forwarded, obtained from my bungalow compound from among many hundred plants growing wild, whose corollas are of different delightful hues.

The plant from which this individual is obtained is *fac simile* to the others, except in the flower, and that its growth is stunted. The uncommon colour— *green*—I venture to say, is doubtless attributable to some chemical change which has taken place in the internal arrangement of this *only plant* from among a number of others. Science teaches us that the leaves of trees and grass, being inclined to be more dark than white, have a greater tendency to absorb than to reflect the solar rays. For instance, the grass and leaves are green, but they absorb all but the green rays. In Professor Henfrey's Botany, 2nd Edition, revised by Dr. Masters, we are told that "the various tints of colour are produced either by the interposition of colourless cells between those containing coloured juices, or by the superposition of cells with different colouring matter one over the other. Then how comes this one plant to be affected more than all the others which are contiguous to it?

The most striking feature in this phenomenon I wish to bring to notice is the abnormal evolution of the corolla having leafy shoots or miniature plants 2 inches high, from whence another flower-bud is shooting—an unheard of *freak*, I think, in this genus! The Honorary Secretary of this section will, I am sure, be glad to explain to us the cause of this metamorphosis, which will be a very interesting lesson to florists who are not versed in teratology.

There are other *green* flowers on the same plant, but at present without any leafy shoots besides the extraordinary one now sent.

F. R.

NOTE ON THE ABOVE,
By Surgeon K. R. Kirtikar.

Mr. Rose's specimen of *Zinnia pauciflora* is an instance of prolification or proliferation, which means the production of one organ by another of a different kind, as that of cup-like appendages by leaves and of branches by flowers or even fruits. For an illustration of this sort of monstrous development, the reader is referred to figures 650 and 774 in Bentley's Botany at pages 286 and 344 respectively (4th Ed.). In the former is an instance of a flower of the Rose showing the axis prolonged beyond the flower and bearing true leaves; in the latter a monstrous Pear has its axis prolonged beyond the fruit and similarly bearing true leaves. In the 5th Edition of Lindley's *Elements of Botany* at p. 62, there is an illustration of the flowers of *Epacris impressa* changing into branches.

This metamorphosis is technically called descending or retrograde where the floral parts, *i.e.*, petals or stamens or carpels become degenerated and are transformed into a leaf. This can be easily explained from the homologous nature of the different parts of a flower to the leaf. A flower in its widest sense is a multiple arrangement of modified or altered leaves. " Linnæus taught it, and Goethe proved it," says Lindley. He mentions an instance from the *Gardener's Chronicle*, in which a Rose is said to have its calyx tube absorbed, at least not manifest; the sepals half converted into leaves; the petals more than half changed into sepals; the stamens fallen off, apparently little changed; the exterior carpels partly in their customary state ; those nearer the centre converted into small leaves; but the remainder upon the axis or centre, which had lengthened into a branch, carried up in every conceivable state of transition, until the last or uppermost carpel assumed the customary appearance of the leaves of the stem. A beautiful illustration is also given by Lindley at p. 63 of his "Elements" above referred to.

The most highly modified leaves of the flower, says Sachs, " are the stamens and carpels." By a freak of nature they may not develop into stamens or carpels, or the stamens and carpels may degenerate into leaves at any time. But though such instances are numerous, they constitute merely a phenomenal transition of an exceptional kind, not necessary for the completion of the life-history of a plant. The floral axis as a general rule ceases to grow at the apex as soon as the sexual organs make their appearance, or even earlier. But in singular or abnormal cases like the one exhibited by Mr. Rose, and *normally in Cycas*, says Sachs, " the apical growth of the floral axis recommences, again produces leaves, and sometimes even a new flower.".

K. R. K.

PROCEEDINGS OF THE SOCIETY DURING
THE QUARTER.

THE usual monthly meeting of this Society took place on Monday the 5th of July, and was largely attended. Dr. D. MacDonald presided.

The following new members were elected :—Mr. E. C. K. Ollivant, Mr. J. A. Betham, Dr. O. H. Channer, Mr. Frank Rose, Mr. F. Chambers, Mr. H. Bromley, Mr. W. J. Holland, Mr. T. B. Fry, Mr. J. H. Steel, Colonel F. W. Major, Mr. Chester Macnaghten, Mr. H. E. Andrewes, Mr. J. Maguire, Mr. G. A. Anderson, Rev. J. E. Abbott, Mr. Cowasji M. Dadabhoy, and Dr. Temuljee B. Nariman. Mr. L. de Niceville, of the Calcutta Museum, was elected an honorary corresponding member of the Society.

Mr. H. M. Phipson, the Honorary Secretary, acknowledged the following contributions to the Society's collections since the last meeting :—

Contribution.	Description.	Contributor.
1 Kestrel (alive)............	From Khandalla	Mr. Wm. Shipp.
1 Sea-snake	Hydrophis curta	Capt. W. P. Kennedy.
1 do.	Pelamis bicolor	Do.
A quantity of Corals ...	From Arabian Coast	Mr. E. H. Aitken.
3 Eggs	Alcippe poiocephala	Do.
A quantity of Birds' Skins and Geological specimens.	From Bhooj	Mr. A. Newnham.
3 Snakes	Zamenis diadema and Echis carinata.	Do.
A quantity of Fossils...	From Beluchistan	Dr. H. Yeld.
A Crow's Nest............	Made of bottling-wire......	Mr. W. M. Macdonald.
1 Snake's Skin (15' 5")..	Ophiophagus elaps	Dr. Bocarro.
3 Floricans' Skins	......	Mr. D. Bennett.
Skin of Pine-marten...	Martes abietum	Capt. Olivier.
3 Ibex's Skins.............	Capra sibirica	Do.
2 Snakes	From Mahableshwar	Mr. J. C. Anderson.
1 Snake (41½')	Cynophis malabaricus ...	Miss Dewar.
1 Elephant's Tooth........	Found at Khandalla	Mr. G. W. Terry.
6 Crocodiles' Eggs (since hatched).	From Tulsi Lake	„ Rienzi Walton, C.E.
2 Snakes	{ Dipsas gokool. Trop. stolatus. }	Lieut. Barnes.
Skin of Albino Mongoose	From Baroda.	Mr. H. Littledale.
2 Young Crocodiles	From Tulsi Lake	„ Nowrojee H. Katrak.
1 do.	......	„ C. A Stuart.
2 Snakes (alive)............	......	„ K. D. Naegamvala.
1 White-tailed Porcupine (alive).	Hystrix leucura	„ A. S. Ritchie.
1 Snake	Trimeresurus anamallensis	„ G. A. Barnett, C.I.E.
1 Camel's Skull	From Sind	„ E. M. Walton.
1 Saw-fish's Snout	Pristis antiquorum.........	Do.
1 Sea-snake	Hydrophis Guntherii	Capt. Fenton.
A quantity of Fish and Crustaceans.	From Alibag	Mr. W. F. Sinclair, C.S.
1 Snake	Gongylophis conicus	Do.
Do.	Onychocephalus acutus ...	Do.
Ratel(or Honey-badger).	Mellivora indica	Mr. R. A. Riddell.
1 Snake	Silybura bicatenata........	„ G. Vidal, C.S.
1 Musang	Paradoxurus musanga......	„ W, F. Hamilton.
1 Booby (alive)	......	„ W. F. Sinclair, C.S.
A quantity of Lizards...	From the Punjab	Major Yerbury, R.E.
1 Snake...................	Zamenis diadema............	Do.
A quantity of Cobra's Eggs.	From Bijapur	Mr. E. Reinhold, C.E.
Head of 4-horned Antelope	......	„ J. D. Inverarity.
A Land Tortoise	From Afghanistan	Major W. J. Morse.
A Rat	Probably a new species ...	Father Dreckman.
2 Snakes	Silybura Elliotii	„ H. E. Andrewes.
Part of a Porpoise's Skull	From Alibag.	„ W. F. Sinclair, C S.
Large Snout of Saw-fish	......	„ Eduljee A. Hormasjee.
1 Python (alive)............	Python molurus	„ H. M. Phipson.

MINOR CONTRIBUTIONS.

From Dr. T. S. Weir, Mr. C. B. Lynch, Mr. H. Curjel, Mr. M. C. Turner, Dr. Dalgado, Mr. F. Jefferson, Mr. E. C. K. Ollivant, Miss Johnston, Mr. F. C. Webb, Mr. A. F. Beaufort, and Mrs. Wright.

CONTRIBUTIONS TO THE LIBRARY.

Transactions of the Linnæan Society of New York, Vols. I. and II.

Records of the Geological Survey of India, Vol. XIX., Nos 1 and 2.

Journal of the Simla Natural History Society, Vol. I., Part 1.

Paper read before the Simla Natural History Society by Colonel H. Collett·

A vote of thanks was then passed to the ladies and gentlemen who had so kindly responded to the request of the Committee and had sent in birds in cages for exhibition at the meeting. The collection consisted of 122 specimens.

Father Dreckman exhibited two full-sized living specimens of the Green Pit-viper (Trimeresurus anamallensis) found at Khandalla, which differed in a very curious way as regards markings and colour.

Mr. Rich also exhibited some beautiful cases of stuffed birds from Australia and New Guinea.

THE usual monthly meeting of this Society was held on Monday 2nd August, Dr. D. MacDonald presided. The following new members were elected :—The Hon'ble F. Forbes Adam, Captain W. P. Kennedy, Mr. S. S. Bengallee, Mrs. Yorke Smith, Mr. F. deBovis, Dr. A. W. F. Street, Mr. P. C. Petit, Rev. H. Juergens, S.J., Mr. F. J. Daley, Mr. E. G. Colvin, C.S., Mrs. A. F. Turner, Khan Bahadur R. J. Ashburner, Mr. Dady M. Limjee, and Mr. Framjee D. Petit.

Mr. H. M. Phipson, the Honorary Secretary, then acknowledged the following contributions to the Society's collection during the past month :—

Contribution.	Description.	Contributor.
1 Crowned-crane.	Ardea pavonina	Victoria Gardens.
1 Chamelion..................	Chamæleo vulgaris	Mr. H. Barrett.
1 Snake	Tropidonotus quincuncia-tus.	Miss Dewar.
A quantity of Tree-crabs	From Mahableshwar	Do.
1 Lizard (alive)	Varanus dracœna............	Mr. W. Killen.
3 Cobras	Naga tripudians	„ H. Littledale.
1 Panther .... ·········	Felis pardus	Victoria Gardens.
1 Wild Ass (Cutch)	Equus onager	Col. Nutt.
4 Snakes	Cerberus rhynchops	Mr. W. F. Sinclair, C. S.
	Gerarda bicolor (?)	Do.
	Onychocephalus acutus ...	Do.
2 Lizards.....................	Varanus dracœna,...........	Do.
A quantity of Crusta-ceans.	From Alibag	Do.
70 Birds	From Ceylon	Mr. A. Newnham.
23 Snakes	Do.	Do.
42 Lizards	Do.	Do.
A quantity of Bactra-chians.	Do.	Do.
A Fœtus of the Mouse-deer.	Do.	Do.
A quantity of Insects.	Do.	Do.

Contribution.	Description.	Contributor.
A quantity of Corals, Shell-fish, and Radiata.	From Ceylon.	Mr. A. Newnham.
A Goat	From Africa	Victoria Gardens.
A quantity of Sea-shells.	From Aden	Mr. J. D. Katelee.
1 Snake.......................	Dipsas gokool	Col. Walcott.
1 Hoopoo	(Mounted in England)......	Capt. Miller.
Specimen of Arraq and Attar.	Made from the Pandanus odoratissimus.	Mr. Frank Rose.
Specimens of Jasper Pudding-stone.	From Banda...................	Do.
1 Hog deer	Axis porcinus	Victoria Gardens.
A quantity of Geological specimens.	From the volcano on Barren Island, Bay of Bengal	Mr. F. J. Daley.
3 Japanese Fish (alive)....	......	Revd. Fr. Dreckman.

MINOR CONTRIBUTIONS.

From Mr. A. F. Turner, Mr. R. MacEwen, Mr. John Fleming, Mr. S. Hedgart, Miss Whitcombe, Mr. A. H. Follet, Mr. J. A. Guider, and Mr. John Dawson.

CONTRIBUTIONS TO THE LIBRARY.

Paper on the Birds of Aden, by Major J. W. Yerbury, R. A.
The Utilization of Minute Life, by Dr. T. L. Phipson.
A Manual on the Diseases of the Elephant, by Mr. J. H. Steel.

EXHIBITS.

A Japanese Dwarf-tree, by Colonel Walcott, and another by the Hon. Mr. Justice Birdwood ; 1 Orchid (in flower) (Phaleonopsis rosea), by Mr. M. C. Turner ; a 4-horned Ram (from Arabia), by Mr. C. E. Kane ; 1 double Cocoanut (from Seychelles Islands), by Mr. A. S. Panday.

THE FUNGI OF BOMBAY.

Surgeon K. R. Kirtikar exhibited a few fresh fungi collected in and round Bombay. The spores of the Bhopud or Lycoperdon or Puff-ball and Hydnum aureatum were exhibited under the microscope, showing the extreme minuteness of the spores of the latter as compared with the spores of the former. Dr. Kirtikar observed that fungi form a very interesting form of plant-life, and, though spoken of somewhat contemptuously as consisting of mushrooms and toadstools, supply the student of Nature with an infatuating subject for observation and amusement. It was a subject, he said, by no means easy of study in this country, especially as previous Indian botanists had paid no special attention to the Cryptogams. Whatever the difficulties, fungi and the other cryptogams, or flowerless plants, afford an interesting field, and would amply repay any trouble that is taken in investigating this unexplored field of some of the most interesting objects in nature. Places around Bombay at this time of the year, when there is so much heat and moisture in the air and in the ground, supply abundant materials for a thorough investigation of this hitherto neglected department of botany. They are not mere toadstools all these fungi, he said, though he showed a tiny toad which he had found sitting on one of the Agarici exhibited. The toads, he said, found not only a stool to sit on, but also a table where they could find their food, as there were numerous earth-worms crawling on the adjacent Polypori. The fungus known to the natives of the country as Phanasamba was a polyporus, and used as a medicinal agent. The Puffball, known as Bhodiphod or Bhopud (*i. e.*, "Cleaver of the soil"), and scientifically known as a variety of Lycoperdon, was, he said, considered a delicacy when properly seasoned and cooked fresh from the field. It appeared on the first fall of the rains in the monsoons. The true mushroom which is sold in the English markets as Agaricus campestris, is also found in this country abundantly, but it is yet too early to find the same just now. It must however be admitted that several pounds of much nutritious food are thrown away as useless on account of want of proper knowledge of the various classes of edible and poisonous fungi. It is not everybody that can relish the musty smell of the varied members of the Fungal tribe, nor is it that the delicacies will always agree with the inner man. But there is hardly any doubt that every student of Nature will find immense delight in scanning the minute threads and spores, and the mycelium or spawn that go to build up the delicate structure of these cellular plants. It is not from the gastronomic point of view that he discoursed, he said, on the fungi, nor was it that he wanted to tell whether this or that mushroom was edible or poisonous, and whether it would do credit to a generous and hospitable host to place before his guest at dinner Indian mushroom toast or stewed or curried toadstools *á lá Indienne !* Nor did he pretend, he said, to initate an energetic mercantile firm into the mysteries of fungus-trade, and encourage a body of speculators to bottle up a few edible varieties and send them to Crosse and Blackwell to try their fate in an English market. All he urged on that evening on behalf of those interesting objects in nature which lie unnoticed was that they had an everlasting interest to the student of Science, and if by such occasional display of fungi the Natural History Society of Bombay encouraged the study of an unexplored field, the Society will have accomplished one of its principal objects. To the student of Medicine the fungi have a special interest, now that fresh accessions are being daily made to our already vast knowledge of bacilli. A thorough acquaintance with their life, history, and their surroundings, and an acquaintance with their habits and functions, are essential before we determine whether they are the cause of disease, concomitants of it, or the mere harmless results of it as any other objects in nature. Fungi of the minutest kind have been known to exist on other larger fungi, apparently not affecting their host with disease or causing its death. Why should not bacilli exist in man without causing disease ? All this has to be known. It was not his intention however, he said, to enter on a medical disquisition, but that he touched the subject incidentally. He then showed from amongst the specimens of fungi some of the typical Agarics, several polypori beautifully tinted, a beautiful golden yellow-spiked Hydnum, the gelatinous ear-like Auricularia, some needle-like Claviarei, thus illustrating one of the important divisions of fungi known as the Hymenomycetes, so called from their possessing the hymenium or fruit-bearing, or rather spore-bearing, surface exposed to the air.

MEMORY AND REASON IN ANIMALS.

Mr. Sterndale then read a paper on "Memory and Reason in Wild Animals." He said that the beginnings of instinct, or he would rather call it reason, began very low down in the scale of animal life, as low down as the Rhizopoda, and he traced it gradually upwards through the Mollusca and Insecta to birds, and from them to the larger animals. He gave some interesting cases of instinct, which he was careful to separate from reason, in monkeys, which rejected certain deadly poisons hurtful to them, and readily took other poisons equally deadly to man and other creatures but which had no effect on them. He then gave cases of reasoning in monkeys and other mammals, and passed on to examples of memory, illustrating his cases by tigers, elephants, horses, &c. He pointed out that in desert islands, untrodden by human foot, there was no instinctive dread of man shown by wild animals. One curious genus of Marine mammalia, Steller's Rhytina, has been exterminated by sailors owing to this over-confidence ; but the advent of man is followed by the loss of this trusting nature, and this is the outcome of reasoning faculties. The wild birds soon see that to confide is to be knocked on the head. Birds, he said, exceeded mammals, with the exception of monkeys, in imitative power. Parrots are made to talk, other birds to whistle. There is no such mimicry amongst mammals, with perhaps the exception of the dog, the bark of which is said to be the unconscious mimicry of the gruffness of the human voice. Wild dogs and wolves cannot bark but only howl. Domestic dogs which run wild lose in a few generations the power of barking and revert to the howl, as in the case of those on the island of Juan Fernandez ; on the other hand, wolf cubs brought up with domesticated dogs learn to bark. He concluded by saying :—" I must however not tax your patience any longer. Did time permit of it, I could give many curious instances of the sagacity of wild animals, their skill in avoiding traps, and their own cunning in circumventing others   The most marvellous creature is the North American wolverine or glutton, regarding which much has been written by Dr. Elliot Coues. I think he heads the list for intelligent rascality, and I recommend such of our members as are interested to turn up the abridged account of it in the second volume of Cassell's Natural History, and they will be amply repaid for five minutes reading. We have nothing like this thoroughpaced villain amongst our comparatively well-behaved denizens of the jungles. I will wind up with a short certificate to his bad character from Dr. Coues :—' The desire for accumulating property seems so deeply implanted in this animal that, like tame ravens, it does not appear to care much what it steals, so that it can exercise its favourite propensity to commit mischief An instance occurred within my own knowledge, in which a hunter and his family, having left their lodge unguarded during their absence, on their return round it completely gutted —the walls were there, but nothing else. Blankets, guns, kettles, axes, cans, knives, and all the other paraphernalia of a trapper's tent, had vanished, and the tracks left by the beast showed who had been the thief. The family set to work, and by carefully following up all his paths, recovered, with some trifling exceptions, the whole of the lost property.' It is well I may say for our Indian police that we have not wolverines among our criminal classes in this country."

The proceedings soon afterwards came to a close.

THE monthly meeting of this Society was held on Monday, the 6th September 1886. Dr. D. MacDonald presiding.

The following new members were elected :—Mrs. John Hay Grant, Captain H. G. E. Swayne, R. E., Mr. K. D. Ghandy, and Mr. S. K. Kambata.

Mr. H. M. Phipson, the Honorary Secretary, then acknowledged the following contributions to the Society's collections during the past month :—

Contribution.	Description.	Contributor.
A quantity of Butterflies	From the Punjab	Major J. W. Yerbury, R.A.
6 Species of Corals.........	From the Mergui Archipelago	Mr. F. J. Daley.
1 Crocodile's Skull	Crocodilus palustris	Do.
Eggs of Monitor	Varanus microlepis.........	Do.
A quantity of Coralines.	....	Do.
1 Snake	Tropidonotus quincunciatus.	Mr. F. A. Montgomery.
A quantity of Bats......	From Oorun	„ E. H Aitken.
14 Snakes	From Saugor, C. P.	„ H. E. Barnes.
Nest and Eggs of the White-eyed Tit.	From Poona	„ R. C. Wroughton.
A quantity of Sponges, Coralines, Crustaceans, Sea-snakes, Fish, and other Marine Animals.	From the Persian Gulf	Capt. E. Bishop.
1 Tree-cat	Paradoxurus musanga......	Mr. J. A. Simpson.
1 Crested Hawk-eagle ...	Spilornis cheela	„ H. Littledale.
1 Snake.........................	Lycodon aulicus	Col. Portman.
22 Birds' Eggs...............	From Ahm-dabad	Capt. F. B. Peile.
1 Markhor's Head	Capra megaceros............	Do.
1 Oorial's Head	Ovis cycloceros	Do.
1 Jungle-cat's Skin.........	Felis chaus	Do.
7 Rats	From Surat	Mr. F. Gleadow.
18 Lizards	Do.	Do.
3 Snakes (alive)	Do.	Do.
6 Rats	Do.	Do.
A quantity of Insects...	Do.	Do.
A quantity of Crustaceans	Do.	Do.
1 Lizard	Eublepharis Hardwickii...	Mr. W. Willard.
1 Snake	Tropidonotus punctulatus.	„ W. F. Sinclair, C. B.
A quantity of Fish and Marine specimens.	From Alibag	Do.
1 Dolphin's Skull	Delphinus plumbeus	Do.
A quantity of Turtles' Eggs.	From Alibag	Do.
A quantity of Sea-shells	Do.	Do.
1 Porpoise, with 2 young ones.	Neomeris kurrachiensis ...	Do.
1 Sea-turtle (alive).........	Covered with Acorn Barnacles.	Do.
A beautifully s t u ff e d specimen of the Duck-billed Platypus.	From Tasmania	Mr. E. M. Walton.
1 Stripe-necked Mongoose.	Herpestes vitticollis	„ N. S. Symons.

MINOR CONTRIBUTIONS.

From Mr. C. W. L. Jackson, Captain Becher, Mr. Thomas Lidbetter, Mr. Mitarachi, Mrs. Owen Dunn, Mr. Forrest, Mr. N. V. Mandlik, Major Morse, Mr. H. Bromley, Mr. C. P. Lynch, Mr. H. Wise, and Mr. Krishnarao V. Ranjit.

CONTRIBUTIONS TO THE LIBRARY.

Records of the Geological Survey of India, Vol. XIX., Part 3.

Insects of India (E. Donovan), by Mr. W. Shipp.

Foreign Butterflies, by Mr. W. Shipp.

Foreign Moths, by Mr. W. Shipp.

Reptiles of Sind (J. Murray), by the Author.

Transactions of the N. S. Wales Linnæan Society, Vol. I, Part 1.

Magazine of Natural History, Vol. XVIII., Nos. 103 and 104, from Mr. W. H. Littledale.

Mr. E. L. Barton exhibited one tiger's head and two panthers' heads, mounted by himself.

The following papers were then read :—

A Matheran Seed-traveller, by Dr. D. MacDonald.

Links in the Mammalian Chain, by Mr. R. A. Sterndale.

Pollen Grains, by Dr. Kirtikar.

A MATHERAN SEED-TRAVELLER.

Dr. MacDonald said :—" Members of the Natural History Society who have visited Matheran in the hot-weather may have noticed seeds with a beautiful crown of spreading hairs—termed *pappus* or *coma* by botanists—carried by the wind, sometimes along the ground, sometimes high in the air. On account of their buoyancy, these wind-wafted seeds are often carried to considerable distances from the parent plant. Several kinds of plants on summit or sides of Matheran Hill produce comose seeds, but perhaps the seeds of which I now show some specimens are the most beautiful. When I first saw these seeds in May of this year, I could not determine their botanical origin, even approximately; but when the then Superintendent of Matheran, Dr. MacDougall, kindly obtained for me some of the leaves of the plant, as well as a few of the maturing fruits, I was able to refer the plant to one of two very closely allied Natural Orders or Families—the *Apocynaceæ*, or Dogbane Order, to which plant so familiar in Bombay as the Allamanda, the Tabernamontana, Vinca rosea, Nerium oleander, Beaumontea grandiflora, and others belong; or the *Asclepiadaceæ* or Milkweed Order, of which the Asclepias curassavica, the Stephanotis, and the Hoya carnosa or Wax-plant are well known in Bombay. These two Orders are very closely allied, and were at one time grouped together under the name Apocynaceæ. The two are now separated, the distinguishing characters of the Asclepiadaceæ being—(1) the *stigma*, which has five rounded angles provided with either cartilaginous corpuscles, or a gland which retains the pollen masses, the stalk or *caudicle* of the pollen masses being attached in this way to the stigma, and (2) the peculiar *pollinia* or pollen masses which are developed by the stamens, instead of the ordinary pollen grains produced by the stamens in the order Apocynaceæ. In the Asclepiads, when the pollen masses adhere to the stigma, the pollen cells simply push the pollen tubes into the lateral and inferior stigmatic surfaces, and thus self-fertilization is effected."

Dr. MacDonald then contrasted the pollen masses found in the Natural Order *Asclepiadaceæ* with those in the Order *Orchidaceæ*, or Orchid Family, in which the pollen masses possess a viscid gland at the base of the stalk or caudicle.

This however was not intended as a means of retaining the pollen masses in the flower in which it was produced, but rather as a means of being carried away to other flowers, as it adhered readily to anything with which it came in contact. As insects were frequent visitors, especially bees and moths, they were often the agents in effecting the cross-fertilization which is the rule in this Order. Pollen masses from the two Orders were shown under microscopes. Returning again to the comose seed, Dr. MacDonald stated that he had identified the plant as the *Anodendron paniculatum* of Dalzell and Gibson's *Bombay Flora*, or the *Gymnema nipalense* of Hooker's *Flora of British India*, the native name being *Lamtani*.

The identification of this plant illustrated the great value of the natural system of classification as compared with the artificial or Linnæan system. The small twig, with its milky juice, the leaves, and the fruit containing the comose seeds, supplied data sufficient to make it certain that the plant belonged to one of two Orders ; and this without the flower, without which any one working on the Linnæan system could not take a single step, as the whole system was based on the parts of the flower.

Dr. MacDonald then pointed out that the stalk of the fruit turned back on itself so as to make the face or side on which it opens turn downwards. As the fruit matures the seeds become loose in the fruit, and when it splits open, as the seeds fall out the wind expands the crown of hairs, and they are thus launched on their voyage of life. The comose crowns, acting as parachutes to prevent the seeds falling at once to the ground, after a time very readily separate, leaving the seeds to germinate where they fall when the rains come.

Before concluding, Dr. MacDonald recommended members who might be interested in the wonderful examples of adaptations of means to an end which occur so frequently in plants to read such books as Sir John Lubbock's recent volume—one of the Nature Series – entitled *Flowers, Fruits, and Leaves*. Although dealing almost exclusively with English plants, any one reading such a book might learn to regard the plants they saw, even in their own compounds, with more interest than hitherto.

LINKS IN THE MAMMALIAN CHAIN.

Mr. Sterndale then read a paper on some links in the Mammalian Chain illustrated by drawings. He said :—" It is common enough to talk of the Animal kingdom as one great chain—and so it is—link is hooked on to link till we find that we are at last the ten billionth cousin of the cabbage we are eating, and so our consciences accuse us of practising homœopathic cannibalism ! You may think this is exaggeration; but look at the *Campanularia*, of which I give here a magnified sketch ; it looks like a plant, it has buds and flowers, and is propagated, we may say, by cuttings, but it is an animal— a Zoophyte—yet how little removed is it in its life from the *Drosera rotundifolia*, the Sun-dew, and other carnivorous plants, which, with surprising life-like attributes, not only catch flies and other insects, but hold them till partly digested.

" These are links which carry us on to our cousin the cabbage, but I do not intend to trace out the pedigree so far. Our time would not permit of such extensive research, so I propose only to give a few of the most curious links in the chain as far as the mammals go. The missing link of course we have not found yet, nor shall we ever find it, for the impassable gulf of intellect separates the brute from man, and no monkey that ever was created will bridge over the gap ; but the barrier of intellect does not operate between ordinary mammals. There have been from time to time abnormal creatures brought forward as missing links, but they have always been human beings with only some monkey-like resemblances.

" Of all those I have seen, the best was a little girl, exhibited about three years ago in London, called *Krao*, or the Missing Link. She was without doubt an ordinary child, very hairy, of the same type as the Burmese family lately exhibited here ; in fact, she came from the same part of the world, not from Burma, but from the adjacent kingdom of Siam. The points dwelt upon were her hairiness, flexibility of her joints : she could lay her fingers back till they touched her fore-arm ; a habit she had of stuffing things into her cheek pouches ; and last but not least the way in which the hair on her fore-arms turned upwards as in the monkeys and not downwards as in men. I looked well at the child, who seemed about six or seven years of age, nd found that she was an ordinary human being, a little more hairy than usual : the flexibility of her hands was merely a matter of training, as also was the habit she had of stuffing her cheeks with grapes, &c. The direction of the hair on her arms was curious, but not in itself sufficient to establish her claim as a link ; so we start from the monkeys. The link between these and the insectivorous animals lies in the Lemurs, which retaining the four hands and some of the anatomical peculiarities of the monkeys, in form and face approximate the carnivorous animals. I have here a living specimen of the mungoose lemur.

" Then from these we go on to the bats. I am not sure that we are anatomically correct in this link, but no other position could be assigned to the flying lemur or *Galeopithicus*.

" The *Galeopithicus volans*, of which I have here a rough sketch, is either a link between the lemurs and the bats or the bats and the insectivora. Naturalists differ on this point. From certain structural peculiarities, I incline to place them before the bats, especially as they are vegetivorous, and therefore should lead on from the lemurs to the frugivorous bats, and not be placed between the insect-eating bats and the insectivora. The animal itself is somewhat like a lemur, but between its limbs it has a membrane exactly like that of the flying squirrel, which I here show you, only that it has this membrane continued round between the hind legs and including the tail as have some genera of bats ; and it is supposed from observations made of its flight that this arrangement enables it to steer itself in its course from tree to tree. In the numerous families of the Order Insectivora there are many curious links, but I have not time to-night to go into them. Anatomically, we must carry on the Insectivora into Carnivora, but talking merely external resemblances, we find much more affinity with the Rodents. Mice and rats are reproduced in shrews : the squirrels are externally like the tupaia. The porcupines have their counterparts in the hedgehogs, and the jerboas in the jumping shrews. A curious instance of similarity is to be found in the squirrels and tupaia. This latter animal is a tree-shrew with a long bushy tail, and when it was first discovered it was considered to be a squirrel till dissection proved it to be an insectivore.

There was subsequently found in the Malayan Peninsula, and I have seen one specimen from Burma, a long-nosed squirrel (*Rhinosciurus tupaoides*), which closely resembles the tupaia.

‘ However, I will bring home to you a still more familiar example in the case of the so-called musk-rat—that most maligned and persecuted little creature which I always encourage in my house, whilst other people destroy it wherever it is found. This miscalled rat is a true shrew, utterly incapable of gnawing a hole through a door or box, and therefore much mischief done by true rats is wrongfully laid to its charge; it comes into your houses for an object which should gain it thanks and protection, and not the violent death it usually meets ; it comes to destroy cockroaches, centipedes, scorpions, and other creeping horrors, and its only offensiveness lies in its powerful odour, which however it only emits when frightened or hurt. I have let one run quietly *five* times over a clean pocket handkerchief without any smell being perceptible afterwards; and the old story of its tainting bottles of beer and wine by simply running over them is a myth. In the old days, when beer and wine were bottled largely in this country, muskratty liquor was common. The bottles were not properly cleaned ; but how seldom do you now hear of the complaint ; it is one of the old Anglo-Indian stories on a par with the cobra in each boot and a scorpion in every keyhole, to say nothing of tigers sitting and licking their lips in the back verandah waiting for the baby ! I have had tame musk-rats and found them them smell less than other pets, certainly not so bad as hedgehogs. At Nagpore a wild one would come out at my call and take grasshoppers from my fingers.

“The most interesting links in the carnivora are those between the cat and dog. The best known is that of the cheeta, of which I have got here a rough sketch ; but he is a true cat, his dentition and internal anatomy are strictly feline, though his claws are not retractile and his form is somewhat dog-like, with long legs and thin body, so he can hardly be called a link. We must go from the cats to the civets and then on from the civets to dogs. There is a curious animal in Madagascar called *Cryptoprocta ferox*, which is a perfect link between the cats and the civets. It is semi-plantigrade, keeping a large portion of the sole of the foot to the ground, and not walking on the tips of its toes like the cats, yet it possesses retractile claws. The skull partakes of the characteristics of both families, and the teeth differ only from the cats in having one more premolar. It is a very savage little creature, muscular and active, and so was appropriately termed *ferox*. The civets are connected with the hyæna by the aard-wolf, a South African animal about the size of a jackall, and in general appearance like a young striped hyæna. It is called *aard* or *earth*-wolf from its habit of burrowing in the ground. The hyænas again are linked on to the dogs by the *Lycaon* or Cape hunting-dog, or hyæna-dog. Here is a rough sketch of one which shews the likeness to both families: the skull is dog-like, but the animal has only four toes on each forefoot instead of five.

“We come now to the Bear family, and we must go back to the cats for a link. No two animals could be more dissimilar than the cat and the bear. Not only are there internal anatomical differences but externally they are unlike the one is light and springy in action, the other heavy and shuffling. The tiger, which is the type of all cats, has but a few sharp cutting teeth which work on each other like a pair of scissors. The bear has more molars, and these with flat crowns, which enable him to grind his food instead of chopping it. The tiger steps lightly on the tips of his toes, with his heel well raised. The bear puts the sole of his foot down flat on the ground. I show you here skeletons of the two animals which will explain what I mean. Now to link the bears with the cats comes a little animal which I have seen in Darjeeling called the *wah* or panda (*Ailurus fulgens*), the red bear-cat. It is bat like in appearace and has semi-retractile claws, but anatomically it is a bear, A larger animal has been found in Eastern Tibet by the Abbe David, and has been called the *Ailuropus*. Only one of these curious creatures has been discovered, and it is still more a link between cat and bear than the other. The *Ailuropus melanoleucos* is about four to five feet in length. The specimen secured measured 4′ 10.″ It is bear-like, as you will see from the rough sketch ; but it is only semi-plantigrade, and its skull exhibits both feline and ursine characteristics; its dentition is feline as regards the premolars, but the true molars are ursine.

"The skull has also a considerable elevation of the occipital crest, and the zygomatic arches are enormous, more so than in any other carnivorous animal : both these are decidedly feline, as you will observe on looking at these skulls of tiger and bear.

"The racoons and the glutton link the bears on to the badgers and weasels, and so on to the otters and sea-otters, and from the last we come to the marine carnivora—the walruses and seals. The sea-otter, though not reckoned among the marine carnivora, is quite as amphibious as a seal ; it is seldom seen on land, though it keeps close in-shore. It has a curious way of floating on its back, and can sleep in that position, and the females do so, holding their little ones between their fore paws.

"From these animals we begin to link on towards the whales. The outward form begins to be fish-like, though the skeleton internally preserves its mammalian character in full ; but the hands and feet lose their grasping powers, and being enclosed in fingerless gloves and stockings, as it were, become mere paddles for swimming. Nothing can be more awkward than a seal or walrus on dry land, yet how graceful in the water. The transition from a seal to a whale or a porpoise is easy to be understood, and here is an argument against the development theory, which is generally underatood to be a progression from a lower to a higher standard. If such transitions take place at all, it would be reasonable to suppose that the porpoise evolved from the seal, for it is not in the fitness of things for a whale or a porpoise to go flopping about on dry rocks till the friction produced legs, whereas we all know that the permanent disuse of any member will lead to its deterioration ; and therefore if we are to have an evolution theory at all, let us suppose that seals took to remaining in the water so long that having no use for legs they left them off. In the cetaceans the upper portion of the skeleton retains the normal mammalian form, but the rest is merely a vertebral column ; hind legs disappear entirely, although the rudiments of small pelvic bones are to be found embedded in the flesh, like the clavicles of the tiger, useless save as a clue.

"Now I have taken up my full share of your time, and have but half gone through my subject. The links between the Rodents, Proboscidea, Ungulata, and Ruminantia must remain over for some future occasion if the subject be deemed of sufficient interest to call for more of it."

POLLEN GRAINS.

Dr. Kirtikar exhibited under the microscope the pollen grains of the Rose hibiscus, Canna indica, Calotropis gigane, Calophyllum inophyllum, Pandanus odoratissimus, Amaryllis, Garuga pinnata, &c., and went on to explain what pollen was. He said it was commonly a yellow powder, sometimes gritty, often impalpable, and was the product of the male portion of the reproductive organs of flowering plants or phanerogams called stamens. It formed an essential element in the process of fertilization or impregnation of the ovule. The pollen of the male organs or stamens and the ovule of the female organs or pistil by themselves, i. e., alone and untouched or unaffected by each other, were powerless in the propagation of the species to which they belonged. The pollen had to come into contact, either directly or indirectly, by being carried from stamens to stigma, from flower to flower, by the busy bee and brilliantly coloured butterflies and moths, or by simple currents of air, winds, and storms. Mr. Blockley's researches have shown that hay fever was caused by the migration of pollen grains of grasses, lilies, roses, and other plants. Professor Otto Thome, of Cologne, the lecturer said, had stated that in forests consisting of those trees which bore catkins, immense clouds of pollen were seen floating in air, at the time of pollination, which were sometimes carried to the earth by showers of rain and there formed the so-called *sulphur-rain*. Special contrivances, Dr. Kirtikar said, existed in water-plants for the utilization of pollen grains. Submerged plants always threw their flower-stalks above the surface of water, as in Trapa sagitta and water-lily. Vallisneria spiralis however had a remarkable mode of fecundation. The male flowers containing the pollen were seated on very short pedicels at the base of the leaves, often several feet below the surface of the water. The female flowers on the contrary had very long pedicels, which at a particular time became greatly elongated and raised the flower to the surface of the water. The male flowers next became detached from their pedicels, rose to the surface, were floated among the female flowers, and thus fertilized the ovule. After this had been accomplished—and this is the most remarkable part of the whole process—the female flower coiled up spirally and the fruit ripened beneath the water. The subject of cross-fertilization which Darwin had so ably followed, the lecturer said, was a study by itself vast and interesting, whereby crossing between different flowers of the same plant, or between flowers on different plants of the same species, was explained.

Pollen, he said, was discharged generally at the time of the opening of the flower, i. e., from the time it completed its bud-state to the time it expanded. The process of pollen-discharge however, he said, might and did continue for some time after the flower had fully opened, but that this happened simply as the remnant of a process which had long since been complete, so far as fructification was concerned, that was to say, that pollen might go on discharging even after the ovule had been acted upon and fecundated. As a general rule, the period of the maturity of pollen and the suitability of the ovule for fertilization were simultaneous. It was noteworthy that in the Natural Orders Orchidaceæ and Asclepiadaceæ, direct fecundation could never take place. In that part of the subject, Dr. Kirtikar said, Dr. MacDonald had anticipated him, and already ably spoken on the subject. An insect must intercede in these orders and transfer the pollinia from one orchid to another.

The pollen cells assumed a variety of forms. Thirty different forms were pictured by Dr. Kirtikar on paper and handed round to the meeting. The contents of the pollen grains, he said, were called fovilla, which consisted of coarsely granular protoplasm containing essential oil and starch globules suspended in finely atomized condition and varying in size from 1-4,000 to 1-30,000 of an inch. It was the essential oil, he said, that gave flowers their value in the world of perfumery.